Environmental Discourses in Science Education

Volume 3

More information about this series at http://www.springer.com/series/11800

Giuliano Reis • Jeff Scott
Editors

International Perspectives on the Theory and Practice of Environmental Education: A Reader

Springer

Editors
Giuliano Reis
Faculty of Education
University of Ottawa
Ottawa, ON, Canada

Jeff Scott
School of Education
Nipissing University
North Bay, ON, Canada

ISSN 2352-7307 ISSN 2352-7315 (electronic)
Environmental Discourses in Science Education
ISBN 978-3-319-67731-6 ISBN 978-3-319-67732-3 (eBook)
https://doi.org/10.1007/978-3-319-67732-3

Library of Congress Control Number: 2017958138

Printed on acid-free paper

This Springer imprint is published by Springer Nature
The registered company is Springer International Publishing AG
The registered company address is: Gewerbestrasse 11, 6330 Cham, Switzerland

Book Abstract

Overall, environmental education (EE) has been aimed at giving people a wider appreciation of the diversity of cultural and environmental systems around them as well as the urge to overcome existing problems. Indeed, universities, schools and community-based organizations struggle to promote sustainable environmental education practices geared towards the development of ecologically literate citizens in light of surmountable challenges of hyperconsumerism, environmental depletion and socioeconomic inequality. The extent that individuals within educational systems are expected to effectively respond to—as well as benefit from—the collective demand for a "greener" and more just world becomes paramount with the vision and analysis of different successes and challenges embodied by EE efforts worldwide. These insights and theoretical perspectives expand the creative commons (i.e. shared social imagination) available for educators worldwide when designing and implementing contextually situated (yet globally relevant and innovative) EE initiatives. In sum, this book shares conceptually and empirically grounded critical perspectives on the conceptualization, implementation, discourses, policies and alternative practices of environmental education for diverse and unique groups of learners in a variety of international educational settings. Each contribution is organized by the various regions of the planet (i.e. continents) and offers insights on the authors' own processes of reimagining an education in/about/for the environment that are realized through their teaching, research and other ways of "doing" EE. In addition, every section provides theoretical and practical scholarship detailing specific frameworks that support and orient research and practices from a multiplicity of angles. This book is intended to foster conversations amongst researchers, teacher educators, schoolteachers and community leaders in order to promote new international collaborations around current and potential forms of environmental education. It is also a means to envision new teaching and research agendas for an education for sustainable development and sustainability.

Endorsements

This book reflects many successful international projects and perspectives on the theory and praxis of environmental education. An eclectic mix of international scholars challenge environmental educators to engage issues of reconciliation of correspondences and difference across regions. In their own ways, authors stimulate critical conversations that seem pivotal for necessary reimaginings of research and pedagogy across the grain of cultural and ecological realities, systematic barriers and reconceptualizations of environmental education. The book is most encouraging in that it works to expand the creative commons for progress in teaching, researching and doing environmental education in desperate times. – **Paul Hart**, Professor of Science and Environmental Education at the University of Regina (Canada), Melanson Award for Outstanding Contributions to Environmental and Outdoor Education (Saskatchewan Outdoor and Environmental Education Association) and North American Association for Environmental Education (NAAEE)'s Jeske Award for Leadership and Service to the Field of EE and Outstanding Contributions to Research in EE.

In an attempt to overcome simplistic and fragmented views of doing environmental education in both formal and informal settings, the collected authors from several countries/continents present a wealth of cultural, social, political, artistic, pedagogical and ethical perspectives that enrich our vision on the theoretical and practical foundations of the field. A remarkable book that I suggest all environmental educators, teacher educators and policy and curricular writers read and present to their students in order to foster dialogue around innovative ways of experiencing an education about/in/for the environment. – **Rute Monteiro**, Professor of Science Education, University of Algarve (Portugal).

Art by: Maria-Luiza Reis

Acknowledgements

We would like to thank the office of the vice-dean of research (Faculty of Education, University of Ottawa) for the invaluable financial support offered to this editorial project during the mandates of Dr. Raymond Leblanc and Dr. Stéphane Lévesque. We extend our thanks to Diandra Singh for her excellent assistance in copy-editing the chapters and Eugenia Vasilopoulos for her work on the original book proposal submitted to the publisher.

We also would like to thank our families for the encouragement provided along the way and for always reminding us that there is more to life than work. Finally, a heartfelt thanks is given to Mike Muller for the vital mentorship dispensed throughout the many steps of the project.

Contents

About the Authors

Wafaa Mohammed Moawad Abd-El-Aal works in science teacher education and environmental education. She is interested in science teachers' pedagogical and subject content knowledge. She has been taking part in many training courses for in-service teachers as well as teaching science in English for pre- and in-service science teachers.

Elizabeth Ashworth is an art education professor in the Schulich School of Education at Nipissing University in North Bay, Ontario. Her research focuses on issues that affect elementary and secondary art education, confidence issues amongst teacher candidates regarding visual arts and embedding environmental considerations within her practice as an art educator.

Jesse K. Butler is a Ph.D. candidate studying educational policy at the University of Ottawa. He specializes in the analysis of public policy issues, particularly as they relate to indigenous and other marginalized communities. His thesis focuses on the enactment of citizenship education policies in urban high schools with high numbers of immigrant and refugee students. Jesse's Ph.D. research is supported by a Joseph-Armand Bombardier Canada Graduate Scholarship.

Qin Chengqiang is the dean of the Sino-Canadian International College of Guangxi University, P.R. of China. His research fields cover applied linguistics, sociolinguistics, cross-cultural communication, environmental education and innovative and entrepreneurship education.

Julie Davis is an adjunct professor of early childhood education at the Queensland University of Technology, Brisbane, Australia . Julie has two areas of research interest: early years' education for sustainability focused on nurturing young children's capacities as agents of change and embedding education for sustainability into teacher education utilizing systems change approaches.

Luiz Marcelo de Carvalho has majored in biology, conducted master's studies on ecology and natural resources and developed Ph.D. research on environmental education trends at primary schools in Brazil. He is currently a professor and researcher at Universidade Estadual Paulista (UNESP) at Rio Claro, Brazil, and coordinates the Environmental Themes and Educational Process Research Group.

Heluane Aparecida Lemos de Souza majored in pedagogy from Universidade Estadual Paulista (UNESP, 2009), has a master's degree in education (2012) and is currently a Ph.D. student in education at the same university. She is also a member of the Environmental Themes and Educational Process Research Group and works as a teacher at a public elementary school.

Sue Elliott is a senior lecturer in the School of Education at the University of New England, Armidale, NSW, Australia . She has been engaged in early childhood education for sustainability advocacy and practice for over two decades. Sue is an acknowledged author in the field and has ongoing research priorities in education for sustainability and forest preschool approaches.

Rita Forte recently completed her Ph.D. in education at the University of Ottawa with a concentration in society, culture and literacies . Her research interests include critical peace education, cosmopolitanisms and the interrelationships between curriculum and governance. One of her favourite experiences was travelling to Morocco to research the spaces of empowerment where students can critically analyse their relations to power.

Mira Freiman works at F.H. Collins Secondary School in Whitehorse, Yukon Territory (Canada). She works as both a counsellor and a teacher, specializing in earth science and geology. In her free time, she adventures in the northern wilderness with her husband Mark Loewenberger.

Myrto Kalliopi Garis is an engineer in profession and currently a Ph.D. candidate on environmental design at the University of Montreal, Canada. With a passion for planning subjects, Myrto works on systems resilience and focuses her research on the long-term effects of reconstruction planning in different time and spatial scales.

Lisa Glithero is a professor with the Faculty of Education at the University of Ottawa. A former teacher, Lisa also served as the education director of Students on Ice, leading global high school youth on over a dozen polar expeditions. Her research interests focus on environmental education, ecocitizenship and the role experiential, nature and place-based pedagogical models play in supporting positive youth development, community youth engagement and democratic citizenship and in addressing global sustainability challenges.

Benjamin C. Herman is an Associate Professor of Science Education at the University of Missouri. His research addresses how sociocultural factors and epistemological beliefs about science and technology impact socioscientific decision-making, particularly regarding environmental issues, and how science education can better prepare people to critically engage in those issues.

Hilary Inwood teaches in the teacher education programmes at the Ontario Institute of Studies in Education (OISE) at the University of Toronto and is the lead on OISE's Environmental and Sustainability Education Initiative. Her research focuses on the professional development of teachers in environmental education, as well as on integrating art education with environmental education to develop environmental literacy in school and community settings.

Ferne McFadden is a special education teacher with the Little Red River Cree School Board in Northern Alberta. In this position, she works to develop programming that supports and challenges students based on individual needs. Ferne believes in inclusive and balanced classrooms and actively works with classroom teachers to find ways that include and support all students.

Joanne Nazir teaches in the bachelor of education programme at Redeemer University College and is a graduate of OISE, University of Toronto. The study of environmental education at Faraway Dale Outdoor Education Centre (referred to in the chapter) was the basis of her doctoral dissertation. Joanne's continuing research interests are in environmental education, science education and teacher education.

Mark H. Newton is an Assistant Professor of Environmental Science Education at California State University, Chico. His research interests focus on the intersection of socioscientific issues, environmental education, and experiential education. Mark holds a Ph.D. in Science Education at the University of South Florida, an M.Ed. in Earth Sciences from Pennsylvania State University and a B.S. in Elementary Education from SUNY Fredonia.

Nicholas Ng-A-Fook is a professor of curriculum studies and director of the teacher education programme at the University of Ottawa. He is the president of the Canadian Society for the Study of Education, the largest professional educational research association in Canada. In these administrative, educational and research capacities, he is committed towards addressing the 94 Calls to Action put forth by the Truth and Reconciliation Commission in partnership with the local indigenous and school board communities. He is collaborating with colleagues to create a state-of-the-art teacher education programme that promises to prepare teacher candidates for the social, economic and cultural demands of the twenty-first century.

Alandeom W. Oliveira is an associate professor of science education at the State University of New York at Albany. His research interests include cooperative science learning, inquiry-based teaching and classroom discourse. He was the recipient of the 2013 NARST Early Career Research Award and 2010 CSSE Distinguished Paper Award.

Patricia Palulis is associate professor in the Faculty of Education at the University of Ottawa. Her research interests involve language, literacy, culture and spatiality. She draws from poststructuralism and postcolonialism and supports the call for decolonizing and indigenizing academia. Currently, she works with eco-literacies and geo-literacies related to children's literature.

Erminia Pedretti is a professor of science education at OISE/UT and teaches in teacher education and graduate programmes, offering courses in environmental education; science, technology, society and environment (STSE); and science education in school and non-school settings. She has received numerous grants to support her research and graduate students. Her current SSHRC grant explores public engagement with science and technology issues through controversial exhibitions in science centres. She recently published two teacher education textbooks for primary/junior and intermediate/senior science methods curriculum and instruction courses. The books are distinctly Canadian drawing on Canadian policies, curriculum and scholarship, with a commitment to nature of science, social justice and environmental education. She has published over 45 articles and 5 books.

Vajiramalie Perera is a primary school teacher in British Columbia, Canada, currently teaching for the Chilliwack School District. Holding a bachelor of science degree (biology) from Simon Fraser University and a bachelor of education degree (intermediate) from the University of British Columbia, Vajiramalie is interested in language, culture and the connections of these to environmental education.

Deborah Prescott majored in biology and anthropology at Whittier College in California and has been an environmental advocate most of her life. She studied her teaching degree externally whilst in Papua New Guinea. Her working life has mostly been teaching EAL/EAD, literacy and numeracy in the Southwest Pacific Islands and Australia .

Giuliano Reis is an associate professor of science education at the Faculty of Education at the University of Ottawa (Canada). His main research interest is in the study of language in/for science and environmental education in various learning contexts. He is married to Juliana Reis and together they have two daughters: Ana-Julia and Maria-Luiza Reis.

Wiske Rotinsulu is a lecturer in the Faculty of Agriculture and postgraduate programmes of Universitas Sam Ratulangi in Manado, Indonesia. Her research interests include geographical information systems and remote sensing in resource management, natural resources management, environmental impact assessment, biodiversity and biosecurity.

Troy D. Sadler is a Professor of Education and the Associate Dean for Research in the School of Education at the University of North Carolina at Greensboro. He conducts research on the teaching and learning of science in the context of socioscientific issues.

Jeff Scott is an Associate Professor in the Schulich School of Education at Nipissing University (Canada). His research program focuses on the intersection of inquiry teaching and learning and Environmental Education with teachers and students in the elementary school system. During recreational time, gardening, canoeing and long distance cycling down quiet country roads occupy his time.

Darren Stanley is associate professor in the Faculty of Education at the University of Windsor. His scholarly research is interdisciplinary; addresses the areas of teaching and learning, curriculum theory and transdisciplinarity; and is framed by theories of complexity and aligned with the themes of epistemology and health .

Astrid Steele works in teacher development in science and environmental education. Her research focuses on teacher development in the overlapping areas of STSE (science, technology, society and environment) and STEM education (science, technology, engineering and math). She is involved in various collaborative projects with colleagues in art and science education.

John Tasirin is a lecturer in the Faculty of Agriculture at Universitas Sam Ratulangi in Manado, Indonesia. He has been involved with the Indonesia Field School since its inception in 2009. John is also a conservation biologist and a member of the Wildlife Conservation Society Indonesia Program and head of a local nature conservation organization in North Sulawesi called PALS. Also an activist, John has worked on the placement of a moratorium on commercial logging in North Sulawesi and to limiting the spread of industrial-scale mining.

Li Tian has worked within New Zealand's community organizations, focusing on the area of volunteering workers' feelings of ownership. His current research is broadly in the area of Chinese tertiary students' well-being.

Feng Yan is an instructor of the Sino-Canadian International College of Guangxi University, P.R. of China. Her research fields cover applied linguistics, translation and environmental education.

Constantinos Yanniris has a decade-long teaching experience in Greece, Egypt and Canada. He is convinced that outdoor environmental education can be effective for the advancement of various educational goals, including environmental literacy. At McGill, his Ph.D. research focuses on the institutional, structural and pedagogical characteristics that constitute efficient environmental education systems.

Xiong Ying is the vice-dean of the Sino-Canadian International College of Guangxi University, P.R. of China. Her research fields cover applied English teaching, cross-cultural communication, environmental education and innovative and entrepreneurship education.

Kelly Young is professor at Trent University's School of Education and Professional Learning. Her research involves studying the effects of a new model of environmental education embedded within an established Eco-Mentorship Certificate Program and alternative teaching field placement, The Learning Garden , through local partnerships for environmental education leadership development.

David Zandvliet is an associate professor in the Faculty of Education at Simon Fraser University in Vancouver, Canada, and the founding director for the Institute for Environmental Learning. An experienced researcher, he has published articles in international journals and presented conference papers on 6 continents and in over 15 countries. His career interests lie in the areas of science and environmental education and learning environments.

Dana L. Zeidler is the Distinguished University Professor of Science Education at the University of South Florida, Tampa. His research program incorporates aspects of argumentation and discourse, moral reasoning and character formation, epistemology and the nature of science into a theoretical framework using socioscientific issues as a means to facilitate scientific literacy.

Chapter 1
Environmental Education: Nurturing a Relationship with Everything, Everywhere

Giuliano Reis, Jeff Scott, and Mira Freiman

Abstract Environmental education (EE) has been deemed capable to re-connect today's youth with the natural environment, thus helping them to become more ecologically responsible citizens. Put differently, EE must be able to challenge people's perceived sense of dis-connection from the planet. This chapter explores ways of nurturing the re-examination of our co-responsibility to life by investigating the content of one course assignment (reading response) submitted by a student teacher and the narratives and actions of a group of young people who embarked on a 10-day canoe trip in Canada. The discussion that follows also serves as a segue into an introduction to the structure of the book.

G. Reis (✉)
Faculty of Education, University of Ottawa, 145 Jean-Jacques Lussier, Ottawa, ON K1N 6N5, Canada
e-mail: greis@uottawa.ca

J. Scott
School of Education, Nipissing University, 100 College Drive, North Bay, ON P1B8L7, Canada
e-mail: jeffs@nipissingu.ca

M. Freiman
F.H. Collins Secondary School, Box 11236, Whitehorse, Yukon Y1A 6N4, Canada
e-mail: mira.freiman@gmail.com

G. Reis, J. Scott (eds.), *International Perspectives on the Theory and Practice of Environmental Education: A Reader*, Environmental Discourses in Science Education 3, https://doi.org/10.1007/978-3-319-67732-3_1

(...)

Laudato sie, mi Signore, cun tutte le tue creature,
spezialmente messer lo frate Sole,
lo quale è iorno, e allumini noi per lui.

Ed ello è bello e radiante cun grande splendore,
de te, Altissimo, porta significazione.

Laudato si, mi Signore, per sora Luna e le Stelle:
in cielo l'hai formate clarite e preziose e belle.

Laudato si, mi Signore, per frate Vento,
e per Aere e Nubilo e Sereno e onne tempo,
per lo quale a le tue creature dai
sustentamento.

Laudato si, mi Signore, per sor Aqua,
la quale è molto utile e umile e preziosa e
casta.

Laudato si, mi Signore, per frate Foco,
per lo quale enn'allumini la nocte:
ed ello è bello e iocondo e robustoso e forte.

Laudato si, mi Signore, per sora nostra madre Terra,
la quale ne sostenta e governa,
e produce diversi fructi con coloriti fiori ed
erba.

(...)

("Il cantico delle creature," *La preghiera del francescano* 2009, p. 789-490)

(...)

Praised be You, my Lord, with all Your creatures,
especially sir brother sun,
who is the day and through whom You give us light.

And he is beautiful and radiant with great splendour,
and bears a likeness of You, Most High One.

Praised be You, my Lord, through sister Moon and
the Stars: in heaven You formed them clear and
precious and beautiful.

Praised be You, my Lord, through brother Wind,
and through the air, cloudy and serene, and every
kind of weather, through whom You give sustenance
to Your creatures.

Praised be You, my Lord, through sister Water,
who is very useful and humble and precious and
chaste.

Praised be You, my Lord, through brother Fire,
through whom You light the night,
and he is beautiful and playful and robust and strong.

Praised be You, my Lord, through our sister Mother
Earth, who sustains and governs us, and who
produces various fruit with colored flowers and
herbs.

(...)

("The canticle of the creatures," *Custodia Terœ Sanctœ*)

1.1 A (Hopeful) Sense of Connectedness

In his "cantico", Saint Francis of Assisi (1181–1221) praises God for all that He had created. More so, he indicates that non-human creatures (i.e. sun, moon, stars, wind, water, fire and earth) are brothers and sisters to us and that all things have one creator and therefore are related. It follows that, as siblings who belong to the same family, God makes no distinction in how He dispenses His love to us and we (everyone and everything) are deserving of the same respect and care. Regardless to one's belief in any existing religious systems or deities, the canticle has a simple message that transcends the boundaries of brick and mortar temples and sacred books: we must look after each other. After all, we are family. Needless to say, our present time couldn't be further from this utopia.

As we write this chapter, the world is knotted up by various socioecological issues, from terrorism to extreme poverty to environmental degradation. For instance, over 65 million people have been forcibly displaced worldwide in 2016 as a result of persecution, conflict, violence or human rights violations (United Nations High Commissioner for Refugees 2017). At the same time, one child in five in high-income countries lives in relative poverty (UNICEF Office of Research 2017), and hundreds of the world's plant and animal species have gone extinct as a result of climate change even though current temperature levels are modest relative to those predicted in the next 100 years (Wiens 2016). In sum, if we were ever commissioned to be kind to one another – human or not – we are failing miserably. Everything, everywhere seems to be on a path towards drastic changes, even extermination.

Although the statistics described above are not improving at any desirable pace, people are becoming more knowledgeable and beginning to make alterations to their lifestyles to help overturn the current state of the planet more than ever before. Not surprisingly, environmental education (EE) has entered the school lexicon (Bodzin et al. 2010), with many now seeking ways to consume less while considering ethically sourced goods (Reis, Mueller, Gisewhite, Siveres and Brito), whereas others work to defy conventional views of happiness that are tied to the mere accumulation of stuff (Diener and Seligman 2004). There are good reasons to get our hopes up (a little bit at least).

This book is but another contribution to these efforts to live in harmony with everything, everywhere. While ambitious in its title, its premise is as simple as the opening canticle: our actions are as much connected as ourselves. So, whatever one of us does to make this world a better place for all can inspire and provoke others within their own confines to provide their contribution. The experiences and ideas shared here might have a rippling effect on readers. At least, that is what we hope.

In the sections that follow, we expand on the significance of being one with the environment and how humans frequently take that affiliation for granted. Next, we present a summary of the different sections under which the chapters selected for the book were placed. In this way, we anticipate that our readers will appreciate the value of this edited collection to both their research and educational practice.

1.2 A Sense of Disconnect

We begin our reflection with a story that happened in a university undergraduate class, which shall remain unidentified for reasons of anonymity. This class had a suggestive general name: Schooling and Society. (In a way, it might imply to some that each one of these phenomena – although intertwined – can be studied separately from one another. Can there be "schooling" without "society"? or "society" without "schooling"? Maybe, but that is beyond the point that we want to make here.) As part of the course, students were required to read an article that studied the feasibility of education for sustainability (EfS) in a specific teacher education programme (see Paige et al. 2016). For one of the assignments, individual students were encouraged to take a more personal approach to the reading by writing a response that considered how the article related to their school/teaching experience. Below is what one student submitted:

> This was an interesting article and although I would've found it more beneficial if it [was] connected to the [provincial]/Canadian context rather than the Australian, there was one particular experience that resonated with me in which I can apply to my own experiences and beliefs as a future educator. Lloyd states that he "always aimed to connect learning to student interests and their local worlds; for example, measuring water quality in the local river or studying geological structures during day and extended excursions." The reason that this particular example stood out to me is because I truly believe one of the most effective ways to engage young learners is to take the inquiry-based approach. With the inquiry-based approach learning is in the student's hands and they are more likely to demonstrate and use higher order thinking skills such as analyzing, evaluating and these skills are way more useful than simply "remembering" textbook information for example. There is so much value when students are in control of their learning and even more value in having young learners out in the environment. I believe having students outside exploring; through the use of their imagination as well as through their curiosity, is one of the most beneficial types of "learning". Children are full of energy, imagination and creativity, and I think its part of my job as an educator to give children the *opportunity* to both explore these facets of their minds and to flourish, in the outdoors and in nature where they aren't confined to walls and artificial lights. When reflecting on my own experience, one of my favourite parts of the day is being on duty during recess. My placement school shares grounds with a city park, meaning that the students have a huge outdoor space to explore. (...) I feel [students] also develop a sense of compassion for nature when they are outside, for example, they learn that they shouldn't break off branches (trees have feelings) and that there are different living things out in the yard and this they connect to both science and religion class.

Although it is unclear the exact reason for the student to believe that the reading would have been "more beneficial" if connected to her immediate context – i.e. not Australian – the same individual was able to pinpoint "one particular experience that resonated with [her]". That is, she nevertheless managed to transcend the locality of the article to imagine how its ideas could be transferred to her condition as a teacher candidate whose placement school "shares grounds with a city park" in Canada. What initially seemed to be unrelatable to her eventually became a possibility. An inquiry-based approach to environmental education is not an exclusive privilege to Australians. Likewise, shouldn't "the *opportunity* to both explore [their imagination and curiosity] and to flourish" be given to students in Canada and everywhere else? Why is that common for people to believe that certain educational issues are exclusive to their immediate realities, thus independent from/unrelated to other circumstances (Reis et al. in press).

In the face of the student's conclusion that she indeed could relate to ideas in the reading, why is that she still decided to indicate that it would have been "more beneficial" to her if the authors had done their study more closely to her known Canadian landscape? In our minds, that student represents those of us who often don't challenge our taken for granted assumptions about the globality/universality of life itself. Once we understand that we are brothers and sisters, naturally entitled to the same rights to a fulfilling communal existence, we will celebrate and respect our similarities rather than use our differences as an excuse for domination, destruction and inaction. But, how can people see the range of our shared current environmental problems and the potential solutions that exist out there? Could "a sense of compassion for nature" develop from just being outside – or "out in the environment"? Outside of what? Of "remembering textbook information"? Of the confines of "walls and artificial lights" of school? Of the streets of our urban lives and inside (city) parks? How can people create *opportunities* (italicized as the student did) to reconsider the connectedness of everything, everywhere? In any way, it appears that people's sense of attachment to their surroundings is fundamental to creating occasions to alter the current social, cultural and ecological (nevertheless embroiled) state of the world.

1.3 Can We "Save" the Planet from Ourselves?

Geological eras are typically characterized by changes in physical systems due to natural events such as volcanoes, glaciation or meteorites. Curiously, the current geological era in which we live – called the Anthropocene (Zalasiewicz et al. 2010) – is characterized by the negative effects that humanity's exploitative behaviours have been having on the planet. Generally, such destructive actions can be associated with a sense of disconnect from the natural environment, which also deludes us from seeing how they indeed threaten our own survival (Weston 2004). Otherwise, for example, although reduced, why would we continue to release chlorofluorocarbon gases in the atmosphere that damage the ozone layer (Crutzen 2006)?

Clayton Barry (2010) suggests that there are four main reasons providing the foundation for the underestimation of our connection with the non-human. Firstly, nature can be seen as mechanistic, controllable and subordinate to humans by virtue of our (self-declared) superior ability to think. This argument is often associated with Rene Descartes, who separated the human mind from everything else (environment) in his famous statement: "cognito, ergo sum" ("I think, therefore I am"). Secondly, nature can be considered an obstacle to human progress. This becomes evident in the language used to discuss our interactions with nature, for example, to conquer a mountain as opposed to climb it. The flood stories in major religions such as Hinduism, Christianity and Judaism could be considered to provide yet another instance of this view (Keay 2000). Thirdly, there is a tendency to distort "'natural laws' to justify human acts of oppression" (Barry 2010, p. 122). That is, our attempts to dominate nature and others of our own kind are but a consequence of an existing natural hierarchy that favours the fittest individuals or groups in society. This is the case when humans see themselves as superior to other beings. Finally, there is the sacred perception of nature, in which humans worship the natural environment (Sudarsana 2013). This too provides a vehicle for separation on the basis that it could lead humans to consider themselves unworthy of being one with a holy natural environment. An example of this type of sociocultural representation can be seen in the movie "The Matrix," where humans are characterized as a disease in a dystrophic world (Wachowski and Wachowski 1999). Although this example is fictional in nature, it reflects the type of environmental symbolism currently present in the North-American popular culture industry and which has been found to continuously influence people's understandings of their own contributions to current environmental issues.

Whatever plausible reasons exist to explain our current state of disconnect from nature, the fact remains that it is deteriorating the wellbeing of the planet (Louv

2008). That is, the development of a healthier relationship with nature is of vital interest to society in general as there is evidence showing the many benefits of connecting with the environment, including but not limited to, an increase in physical fitness, creativity, self-esteem, attention restoration and reduced stress (Charles et al. 2009). However, only "when we see land as a community to which we belong, we may begin to use it with love and respect" (Leopold 1949, p. xvii). (This is exactly what St. Francis' canticle suggested in the thirteenth century.) In this scenario, could EE help to strengthen our connections with everything else around us, human or otherwise? Do schools still hold the potential to help us in this process of reconnecting?

1.4 Can Schools Help Us "Save" the Planet from Ourselves?

As an organization, the formal school system has been deemed to help build the necessary knowledge and attitudes to support a closer connection with the environment, especially through a historical connection with science (Campbell and Robottom 2004). Contradictory as it may seem, current school practices can also reinforce a disconnection from nature. In a reality regulated by inexact educational policies and strong socioeconomical inequalities (Wall 2000), the existence of educational disciplines that separate social, political, economic and environmental knowledge makes it difficult for students to understand nature and human society as anything but separate from one another (Barry 2010). Indeed, the "institutionalization [of environmental education] within general education works against its own socially and ecologically transformative goals" (Gruenewald 2004, p. 71). As a result, teachers and students are often challenged to promote and advance EE programmes that are short-lived (Reis and Guimaraes-Iosif 2012). That is, "the complexity of interactions which determine behaviour [or action], illustrates that environmental citizens are not produced merely by programmes of education, but by a whole range of factors which education may interact" (Hawthorne and Alabaster 1999, p. 40).

In the context where formal education is not equipped to solely undertake the transformative role needed to face current environmental challenges, environmental educators in all learning settings are challenged to find alternative strategies to reconnect humans with nature. One possible solution would be the direct exposure to nature through outdoor recreational activities (Hanna 1995). Thus, the integration of field experiences into the school curricula could benefit teachers and students in restoring their relationship with the environment more than traditional classroom-based instruction alone (Dettmann-Easler and Pease 1999). We now turn to the examination of one such field activity: canoeing. It is meant to serve as an analogy for the type of journey that we hope our readers will embark while reading this book, one that aims to restore our (presently debilitated) interdependence with everything, everywhere.

1.4.1 Canoeing at Deep Waters

Camp Deep Waters is a privately run co-ed summer camp for youth aged 7–18, founded in 1931. It provides a diversity of canoe tripping programmes that range from 1 to 56 days. As a member of the Ontario Camping Association (OCA), the camp conforms to specific standard policies. For example, its trips must follow low-impact environmental practices, which are meant to minimize human disturbance to visited locations (Ontario Camps Association 2008). It must also adhere to safety protocols, such as the one that requires individuals to wear lifejackets when paddling on flatwater to helmets and lifejackets while paddling a white-water rapid.

At Camp Deep Waters, Mira participated in a 10-day canoe trip with 11 other people: two adults (a 20-year-old male and a 23-year-old female) and nine youths (five females and four males, ages ranging from 13 to 15 years). During that time, participants were encouraged to produce artefacts (drawings and photographs) representing their connection with/disconnection from nature. Additionally, Mira took notes on participants' accounts of activities and other observable events.

The first 2 days of the canoe trip were spent in nature in an area with fewer people and different amenities than one would expect to see in a heavily urbanized area: there were lakes, tress, canoes and paddles, but no sight of tall buildings, long sidewalks or malls. The third day of the trip involved passing through a town, which was perceived by most participants as a typical symbol of civilization. Days 3–9 were spent paddling down a river in remote surroundings. Electrical wires, roads and houses were to be found again only nearing the end of the trip on day 9 (pseudonyms are used throughout to preserve anonymity of participants).

1.4.2 Nature-Civilization Continuum

Markedly, the factor that most clearly mediated participants' sense of connectedness to nature during the trip was the variation in the intensity with which the presence of other humans was perceived in the environment. For instance, one of the young members of the group (Hawk) expressed feeling connected with nature as he stood in the rain on the fifth morning of the trip: "[the rain] washed away all the human smell and made all the air smell better, like nature – the trees, the leaves and the water". In this case, Hawk's experience with the surroundings is prominently embodied (kinaesthetic) (Reis 2015).

In addition, a number of artefacts created by participants provided further support to the idea of feeling connected with the environment when human presence is minimal. This was the case when Otter, another young participant, drew an illustration of a sapling (Fig. 1.1) and explained that it was linked to the rocks, moss and river. Curiously, his observation of nature led him to ascertain that everything in nature is connected even though humans are nowhere to be seen in the drawing.

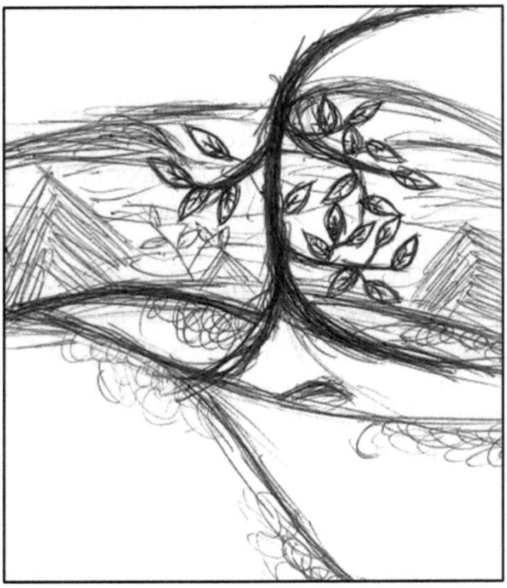

Fig. 1.1 Otter's depiction of the sapling connecting with the surrounding environment (© 2017 Mira Freiman)

Another example that illustrates participants' perceived sense of connectedness to nature could be seen in their playful interactions with insects, like when a coyote chatted with mayflies that were on his arm or when a bear greeted a large beetle on the canoe barrel (though he was not sure if it was a dangerous or safe beetle): "Whoa, hello big boy". (At around the same time, a wolf went after it with his paddle despite the requests from two other youths not to kill it.)

Mira's notes (edited here for clarity and brevity) reveal other remarks made by participants about the impact that indications of human presence had on them during specific parts of the trip:

> There are motorboats flipped over on the rocky shoreline. I can see dead fish in the water and what appear to be pieces of an old wood-canvas canoe. Otter says: "This place makes me feel further from nature." I ask "why?" and she says something about it being "creepy." Deer agrees with her. Fox also says that the area makes her feel further from nature. (Day 2, 4–5pm)

> Otter says being here makes her feel more connected to nature. I ask "really?" and she confirms "yeah." Loon says that being here makes her feel further away from nature. Otter explains that it is beautiful here and she likes looking out and seeing the view. Deer and Otter agree that stepping back from nature makes them feel more connected to nature. I am reminded of the saying 'distance makes the heart grow fonder', which implies in this context that distancing from nature helps us connect with nature, and appreciate it more. Bear asks if it is like how when you are in the city, you do not appreciate how comfortable your couch is. There is a chorus of agreement. Otter says when she is in civilization, she wants the woods, and when she is in the woods, she wants civilization thereby suggesting that the woods and civilization are mutually exclusive. (Day 3, 12pm)

Other human activities typical to the outdoors were also considered to influence participants' perception of their connectedness to nature. That was seen on the morning of day 3, when Fox took a picture of Robin making breakfast and said that it made her feel closer to nature.

Generally speaking, despite the fact that perceptions of connectedness to nature differed amongst individual participants (i.e. some felt discomfort while standing in the rain while others enjoyed it), an object or sight perceived to be associated with the natural setting of the canoe trip was most likely than not to make participants feel closer to nature (e.g. rain, insects, trip canoes and group members). Conversely, if something was perceived to belong outside of the canoe trip (e.g. abandoned old wood-canvas canoe and signs of urbanization), it had the potential to make participants feel disconnected from the same natural environment. In sum, participants' sense of closeness to/detachment from their surroundings fluctuated along the canoe trip according to their interpretation of what they saw, heard, tasted, smelled and touched during those 10 days.

The distinctive meanings of nature and civilization suggested by participants' narratives and actions indicate that they are aware of/triggered by the particularities of each situation. In the end, the canoe trip exposed the youth to different circumstances along a nature-civilization continuum. In creating possibilities for people to reflect on the elements that can alter their perceptions of proximity to – or distance from – nature, one can only hope to empower the individual to develop a more holistic notion of nature, where our humanness is but an intrinsic (and perhaps small) part of life on the planet. And this is one of the ambitious goals of this book: to invite readers to consider different (but real) possibilities to bring themselves closer to nature.

1.5 This Book

The present introductory chapter discussed the pedagogical aspects of a typical teacher education assignment, a reading response, and a particular EE initiative, a canoe trip, as their implications to people's sense of connectedness to nature. In the first story, the Canadian teacher candidate initially failed to see how her educational context was not so distinct from that of Australia's. She felt disconnected from that other school system even though she was able to draw parallels between the two in her assignment. Likewise, in the second story, canoe trip participants were met with distinct feelings about their sense of connectedness to nature during their excursion into the wild. In common, the two stories have the fact that the people involved reflected on their relationship to nature (i.e. moments when they appeared to be close to/distant from their surroundings). In doing so, they were able to identify those elements that helped them to feel less distinct from their environment and challenged their taken for granted assumptions about how connected/related they really are to one another – human or not. The present book has been written to provoke this type of conversation/reaction/reflection amongst readers. How can we get closer to the environment? How can humans restore and strengthen a respectful relationship with others as well as with themselves? How can we develop a greater sense of "compassion" for what is around us and for ourselves? Are there different ways we can do this? If so, would they be unique to specific geographies or could they be adapted to different locations on the map? How can we – whoever and wherever you are – create occasions to alter the current social, cultural and ecological (nevertheless embroiled) state of the world? You will find possible answers to these questions here.

The authors represent a diversity of places and perspectives on EE. Given the potential high number of scholars who could have contributed to the book, contributions were selected based on their critical aspect and the fact that they were not necessarily bounded by their geographies. That is, in selecting the chapters that you now see here, preference was given to those that were truly – or as much as possible – international in the sense that readers from elsewhere could clearly relate to their messages. For most of us educators, the complaint that an idea or example is inadequate due to its lack of local applicability can be sometimes frustrating. Of course, one could not expect environmental matters to be approached or even valued equally around the planet and without any consideration for the original contexts where they emerged. Problems and their solutions are never universal (i.e. one size fits all). At the same time, one idea in Australia might inspire/assist people in Mexico (or vice versa) to continue on with their efforts of making the world a more just place for humans and non-humans. This is the international (perhaps global) aspect of each chapter.

The contributions in the book have been placed under four interrelated sections. The first one is called "Environmental Education and Teacher Education" and includes contributions more strongly correlated to the theory and practice of EE within teacher education programmes:

- *Understanding ecojustice education as a praxis of environmental reconciliation: Teacher education, indigenous knowledges, and relationality* (Chap. 2) by Jesse Butler, Nicholas Ng-A-Fook, Rita Forte, Ferne McFadden and Giuliano Reis (University of Ottawa). In this chapter, the authors apply the concept of reconciliation to environmental education (EE), exploring principles through which EE scholars, both in Canada and internationally, can take up EE as a praxis of environmental reconciliation.
- *Reimagining environmental education as artistic practice* (Chap. 3) by Hillary Inwood (University of Toronto) and Elizabeth Ashworth (Nipissing University). Here, the authors explore the potential of art education to increase the power and reach of environmental learning in teacher education programmes and schools. Likewise, the concepts of creativity, imagination and aesthetics are considered powerful pedagogical tools to provide a means for teacher educators of the arts to contribute towards positive environmental change.
- *Integration, inquiry, and interpretation: A learning garden alternative placement and eco-mentorship program for pre-service teachers* (Chap. 4) by Kelly Young (Trent University) and Darren Stanley (University of Windsor). In this chapter, the authors address the complex nature of environmental education by highlighting some of the ways in which language – specifically, root metaphors – perpetuate the ongoing disconnection of humans and the natural world.
- *Embedding environmental sustainability in a predominantly online teacher education program: ways to contextualize learning* (Chap. 5) by Deborah Prescott (Charles Darwin University). Through a teacher self-study, this chapter illustrates

aspects of learning and assessment design in the context of a regional Australian tertiary institution with predominantly online delivery of two large courses for teacher candidates. The four key guidelines for effective learning and assessment design that emerged through the investigation are considered fundamental to effective environmental education programmes and cross-disciplinary learning.

The second section of the book is titled "Environmental Education outside Walls" and contains those chapters that speak to the theory and practice of environmental education in nonschool settings:

- *Educating for student agency: Perspectives from young eco-civic leaders in Canada* (Chap. 6) by Lisa Glithero (University Of Ottawa). This chapter shares the findings of a study that investigated how youth, nationally recognized as eco-civic leaders in Canada, perceive their agency and capacity to create a more environmentally sound society.
- *Environmental education as/for environmental consciousness raising: Insights from an Ontario outdoor education centre* (Chap. 7) by Joanne Nazir (Redeemer University College) and Erminia Pedretti (OISE, University of Toronto). Here, the authors draw from insights gained from a study of educators working together at an outdoor education centre in Ontario (Canada) to advance the idea that raising environmental consciousness involves connecting people to, fostering care for and building agency for their environments.
- *Indonesian adventures: Developing an ecology of place on Sulawesi Utara* (Chap. 8) by Vajiramalie Perera (Simon Fraser University), Wiske Rotinsulu (Universitas Sam Ratulangi), John Tasirin (Universitas Sam Ratulangi) and David Zandvliet (Simon Fraser University). In this chapter, the authors develop an island metaphor and explore the meaning of community in place-based education.

The third section of the book is called "Environmental Education in the Context of Schools" and includes those contributions that are built upon the notion that schools can be critical venues for environmental education to take place:

- *Crisis and recovery in environmental education: The case of Greece* (Chap. 9) by Constantinos Yanniris (McGill University) and Myrto Kalliopi Garis (Université de Montréal). This chapter explores the institutional, structural and pedagogical characteristics that contribute to the resilience of environmental education frameworks towards the survival of future socioeconomic perturbations in a world of growing uncertainty.
- *Storied environmental curriculum: A case-based perspective on environmental education* (Chap. 10) by Alandeom W. Oliveira (State University of New York at Albany). Here, the author examines the written organization of two cases developed by biology teachers in Brazil to highlight the need for environmental educators to better understand what it means to develop a storied curriculum on ecological issues.
- *A socioscientific issues approach to environmental education* (Chap. 11) by Benjamin C. Herman (University of Missouri), Troy D. Sadler (University of Missouri), Dana L. Zeidler (University of South Florida) and Mark Newton

(California State University). This chapter makes a case for applying a socioscientific issue (SSI) approach to environmental education (EE).
- *Moving forward from the margins: Education for sustainability in Australian early childhood contexts* (Chap. 12) by Sue Elliott (University of New England) and Julie Davis (Queensland University of Technology). The authors of this chapter offer an analysis of early childhood education for sustainability (ECEfS) theory and practice in Australia as means to ramp up ECEfS activities elsewhere.
- *Environmental education in China: A case study of four elementary and secondary schools* (Chap. 13) by Qin Chengqiang, Xiong Ying, Feng Yan and Li Tian (Guangxi University). This chapter is based on a survey conducted within four elementary and secondary schools in Nanning (China) and discusses important questions related to the improvement of environmental education in these schools and recommendations for others.

The fourth and final section of the book – "Environmental Education Research and Poetry" – is dedicated to those contributions that are focused on the scholarship of environmental education but with a more personal twist:

- *An improbable international collaboration: Finding common ground* (Chap. 14) by Astrid Steele (Nipissing University) and Wafaa M. Abd-El-Aal (Beni-Suef University). Here, the authors share the circumstances and development of their initial collaborative environmental education project while also reflecting on the benefits and obstacles of their continued partnership.
- *Environmental education research and the political dimension of education for citizenship: The Brazilian context* (Chap. 15) by Luiz Marcelo de Carvalho and Heluane Aparecida Lemos de Souza (Universidade Estadual Paulista). In this chapter, the authors explore the relationship between education and the process of citizenship construction in environment education research reports and theses carried out in Brazil in 1981 to discuss the dangers of universalizations in the field.
- *Black earth green moon Mama Allpa: Polyphonic moments from temple to tambo* (Chap. 16) by Pat Palulis (University of Ottawa). Drawing into the concept of ecopedagogical responsibility, the author makes a poetical stand against the grain of the progress narrative.

Similarly to a student who is introduced to new and exciting ideas in school or a canoe trip participant who has a chance to explore first-hand the outdoors, the editors would like to invite readers to embark on a journey when reading the book. To that end, each chapter has also been written with the goal of increasing people's likelihood to buy into EE and therefore feel empowered to act upon the possibilities that exist within the theory-practice realm of the field. That required all of us – authors and editors – to depart from at least three convictions: (a) that learning is contextually driven by a myriad of social, cultural, personal, affective and physical elements (Falk and Dierking 2000); (b) that in order for learners to reconnect with nature, they must act, think and feel with/in it; and (c) that the natural environment can indeed be a teacher in and of itself (Wals 2012). We ask that our readers too be open to these principles.

Finally, we hope that the content of this book will be considered for use in undergraduate and graduate university programmes worldwide. As such, authors provided discussion questions aimed to help instructors and students to engage in meaningful conversation about the ideas presented. These questions are located at the end of each chapter and by all means should be utilized and modified as needed. Enjoy!

Questions

Discussion questions that could help instructors and students to engage in meaningful conversation about the ideas presented in this chapter.

1. In your opinion, what does it mean to be disconnected from nature? Please provide two examples to illustrate your answer
2. What is "the environment"? How is it different or not from your understanding of "nature"?
3. Indicate one local socioecological issue in your community. Do you believe that this problem is exclusive to your area or it is common in other countries? (You will have to do a quick online search to support your answer here)
4. Have you ever been on a canoe trip or a similar outdoor experience? Were your emotions similar to those exhibited by the canoe trip participants? Please explain your answer
5. Do you see signs of human interference in the area where you live? How do they make you feel: connected or disconnected from the natural environment? Why?
6. In your opinion, how could formal educational settings contribute to making people feel closer to nature?

References

Barry, C. (2010). The environment/society disconnect: An overview of a concept tetrad of environment. *The Journal of Environmental Education, 41*(2), 116–132. https://doi.org/10.1080/00958960903295241.

Bodzin, A., Klein, B., & Weaver, S. (2010). *The inclusion of environmental education in science teacher education*. New York: Springer.

Campbell, C., & Robottom, I. (2004). Environmental education: Appropriate vehicle for science education? *Teaching Science, 50*(2), 18–23.

Charles, C., Louv, R., Bodner, L., Guns, B., & Stahl, D. (2009). *Children and nature 2009: A report on the movement to reconnect children to the natural world*. Santa Fe: Children & Nature Network.

Crutzen, P. (2006). The "anthropocene". In E. Ehlers & T. Krafft (Eds.), *Earth system science in the anthropocene* (pp. 13–18). New York: Springer.

Custodia Teræ Sanctæ. (n.d.). *The canticle of the creatures*. Available at http://www.custodia.org/default.asp?id=1454

Dettmann-Easler, D., & Pease, J. (1999). Evaluating the effectiveness of residential environmental education programs in fostering positive attitudes toward wildlife. *Journal of Environmental Education, 31*(1), 33–39. https://doi.org/10.1080/00958969909598630.

Diener, E., & Seligman, M. (2004). Beyond money: Toward an economy of well-being. *Psychological Science in the Public Interest, 5*(1), 1–31. https://doi.org/10.1007/978-90-481-2350-6_9.

Falk, J., & Dierking, L. (2000). *Learning from museums: Visitor experiences and the making of meaning*. Walnut Creek: AltaMira.

Gruenewald, D. A. (2004). A Foucauldian analysis of environmental education: Toward the socioecological challenge of the earth charter. *Curriculum Inquiry, 34*(1), 71–107. https://doi.org/10.1111/j.1467-873x.2004.00281.x.

Hanna, H. (1995). Wilderness-related environmental outcomes of adventure and ecology education programming. *Journal of Environmental Education, 27*(1), 21–32. https://doi.org/10.1080/00958964.1995.9941968.

Hawthorne, M., & Alabaster, T. (1999). Citizen 2000: Development of a model of environmental citizenship. *Global Environmental Change, 9*(1), 25–43. https://doi.org/10.1016/s0959-3780(98)00022-3.

Keay, J. (2000). *India: A history*. New York: HarperCollins.

La preghiera del francescano. (2009). Padova: Messaggero di Sant'Antonio.

Leopold, A. (1949). *A Sand County almanac and sketches here and there*. Oxford/New York: Oxford University Press.

Louv, R. (2008). *Last child in the woods*. New York: Workman Publishing Company.

Ontario Camps Association. (2008). Trips and excursions: Relevant regulations and resources. *Guidelines for Accreditation*. Available at http://www.ontariocamps.ca/documents/parents/OCA%20Guidelines%20for%20Accreditation%202008.pdf

Paige, K., Lloyd, D., & Smith, R. (2016). Pathway to 'knowing places' – And ecojustice – Three teacher educators' experiences. *Australian Journal of Environmental Education, 32*(3), 260–287. https://doi.org/10.1017/aee.2016.18.

Reis, G. (2015). A socio-culturally sensitive science curriculum: What does it have to do with our bodies? In C. Milne, K. Tobin, & D. Degenero (Eds.), *Sociocultural studies and implications for science education: The experiential and the virtual* (pp. 231–255). New York: Springer.

Reis, G. (in press). Science teachers under suspicion: Is it true that science teachers aren't good as they used to be? In L. Bryann, K. Tobin (Eds.), *13 questions in science education*. New York: Peter Lang.

Reis, G., & Guimaraes-Iosif, R. (2012). The death and life of a school-based environmental education and communication program in Brazil: Rethinking educational leadership and ecological learning. *Applied Environmental Education and Communication, 11*(3–4), 123–132.

Reis, G., Mueller, M., Gisewhite, R., Siveres, L., & Brito, R. (in press). *Sociocultural perspectives on youth ethical consumerism*. New York: Springer.

Sudarsana, C. (2013). An empirical study on the relationship between nature worship and the preservation of environment in some parts of West Bengal in India. *International Journal of Social Sciences, 2*(2), 79–83. https://doi.org/10.5958/j.2321-5771.2.2.002.

UNICEF Office of Research. (2017). *Building the future: Children and the sustainable development goals in rich countries, n. 14*. Florence: UNICEF Office of Research- Innocenti. Available at https://www.unicef-irc.org/publications/series/16/.

United Nations High Commissioner for Refugees. (2017). *Global trends 2016*. Geneva: UNHCR. Available at www.unhcr.org/statistics.

Wachowski, A., & Wachowski, L. (1999). *The matrix [motion picture]*. United States: Groucho II Film Partnership, Silver Pictures, Village Roadshow Pictures, and Warner Brothers Pictures.

Wall, G. (2000). Recent representations in popular environmental discourse: Individualism, wastefulness and the global economy. *Canadian Review of Sociology and Anthropology, 37*(3), 249–265.

Wals, A. (2012). Learning our way out of unsustainability: The role of environmental education. In S. Clayton (Ed.), *The Oxford handbook of environmental and conservation psychology* (pp. 628–644). New York: Oxford University Press. https://doi.org/10.1093/oxfordhb/9780199733026.013.0032.

Weston, A. (2004). What if teaching went wild? *Canadian Journal of Environmental Education, 9*, 31–46.

Wiens, J. (2016). Climate-related local extinctions are already widespread among plant and animal species. *PLoS Biology, 14*(12), e2001104. Available http://journals.plos.org/plosbiology/article?id=10.1371/journal.pbio.2001104. doi:10.1371/journal.pbio.2001104.

Zalasiewicz, J., Williams, M., Steffen, W., & Crutzen, P. (2010). The new world of the anthropocene. *Environmental Science and Technology, 44*(7), 2228–2231. https://doi.org/10.1021/es903118j.

Part I
Environmental Education and Teacher Education

Chapter 2
Understanding Ecojustice Education as a Praxis of Environmental Reconciliation: Teacher Education, Indigenous Knowledges, and Relationality

Jesse K. Butler, Nicholas Ng-A-Fook, Rita Forte, Ferne McFadden, and Giuliano Reis

> *Unless Indigenous and non–Indigenous students understand Indigenous world view and values, they will never be effective change agents toward healthy cross–cultural relations.*
>
> Bell 2011, p. 383

Abstract In recent years, reconciliation has become a central concept in renewing relationships between Indigenous and non-Indigenous communities in Canada. In this chapter, we apply this concept to environmental education (EE), exploring principles through which EE scholars, both in Canada and internationally, can take up EE as a praxis of environmental reconciliation. In particular, we analyze the literature on ecojustice education, discussing both the possibilities and the limitations of this framework in relation to Indigenous education. We then present qualitative findings from teacher candidates (TCs) completing a voluntary practicum in an Indigenous community and discuss how the findings indicate the shortcomings of

J.K. Butler (✉)
University of Ottawa, 136 Jean-Jacques Lussier (5002E), Ottawa K1N 6N5, ON, Canada
e-mail: jessekbutler@gmail.com

N. Ng-A-Fook • G. Reis
University of Ottawa, 145 Jean-Jacques Lussier, Ottawa K1N 6N5, ON, Canada
e-mail: nngafook@uottawa.ca; greis@uottawa.ca

R. Forte
University of Ottawa, Ottawa, Canada
e-mail: rfort017@uottawa.ca

F. McFadden
University of Ottawa, P.O. Box 120, John D'or Prairie T0H 3X0, AB, Canada
e-mail: fernemcfadden@gmail.com

G. Reis, J. Scott (eds.), *International Perspectives on the Theory and Practice of Environmental Education: A Reader*, Environmental Discourses in Science Education 3, https://doi.org/10.1007/978-3-319-67732-3_2

current teacher education practices in relation to EE and the need for an environmental reconciliation-oriented approach. Finally, we provide specific recommendations for EE scholars elsewhere who wish to take up EE as a praxis of environmental reconciliation.

In 2007, a group of Canadian Indian Residential School survivors won a class action settlement agreement against the Government of Canada, worth an estimated 2 billion dollars. One of the outcomes of this settlement was the formation of the Truth and Reconciliation Commission (TRC) of Canada, with a mandate to uncover the history of Indian Residential Schools in Canada and to foster reconciliation between Indigenous and non-Indigenous communities. Between 2008 and 2015, the TRC traveled across Canada, listening to the stories of survivors and facilitating various public commemorative events. Now, with the conclusion of the TRC's mandate, reconciliation has become an important concept when discussing the relations among and between Indigenous and non-Indigenous Canadians. And yet, as Jennifer Henderson and Pauline Wakeham (2013) have pointed out, the common use of the word "reconciliation" among various parties can "obscure the complex negotiations surrounding the slippage of meanings attributed to this word" (p. 9). The underlying question, according to the same authors, is what degree of change to the status quo is required for reconciliation to take place. In a settler colonial state such as Canada, one central aspect of the status quo that stands in the way of reconciliation is the relationship of the people to the land they live on (Veracini 2010). Canada was founded on the forcible displacement of Indigenous peoples from the land they had lived on since time immemorial (McCrossan 2015). During this process of colonization, Indigenous peoples' traditional reciprocal and holistic relationship to their environment was replaced with an industrial model for envisioning the environment as an object and a resource existing apart from an artificially abstracted society (Henderson 2000).

Now that the TRC's official report has been released, we believe it provides important ways forward toward meaningful social, political, and environmental reconciliation in Canada. In particular, the TRC (2015) puts forward an important understanding of "reconciliation as relationship" (p. 21). This understanding holds an important potential for renewing environmental education (EE) as a situated relationship with local communities and their respective environments. According to the TRC:

> Reconciliation between Aboriginal and non–Aboriginal Canadians, from an Aboriginal perspective, also requires reconciliation with the natural world. If human beings resolve problems between themselves but continue to destroy the natural world, then reconciliation remains incomplete. This is a perspective that we as Commissioners have repeatedly heard: that reconciliation will never occur unless we are also reconciled with the earth. Mi'kmaq and other Indigenous laws stress that humans must journey through life in conversation and negotiation with all creation. Reciprocity and mutual respect help sustain our survival. (p. 18)

The TRC here suggests a clear way forward for EE in Canada, through renewing local relationships between the community and the environment. However, in many Indigenous communities in Canada, this local and relational approach toward education remains complicated by the high number of non-Indigenous teachers from outside the community, often from distant urban centers, living and teaching on their reserve (Butler 2016).

In response to this dislocating situation, this chapter examines the process of non-Indigenous teacher educators and candidates learning to teach EE in a culturally relational manner within an Indigenous community. We seek to answer the following question: How can non-Indigenous teachers coming from outside an Indigenous community engage in EE in relational ways that honor local Indigenous knowledges? In doing so, we also hope to deepen our understanding of the ways in which the Teacher Education Program at the University of Ottawa continues to reproduce a neocolonial discursive regime that excludes local Indigenous knowledges (Butler et al. 2015). Such exclusions, we argue, create various epistemological and discursive barriers for TCs (and university educators) who are committed toward creating spaces to develop and become ecologically literate citizens.

In this chapter, drawing on our experience with an ongoing collaborative partnership between a Kitigan Zibi Algonquin community and the Faculty of Education at the University of Ottawa, we seek to understand the potential relations between EE and reconciliation. In each year since 2011, 10–15 TCs, enrolled in the Developing a Global Perspectives (DGPE) cohort within the Teacher Education Program at the University of Ottawa, have been working with First Nation teachers and students at the Kikinamadinan elementary and secondary school on the Kitigan Zibi reserve, located 90 min north of our settler capital, in Quebec, Canada. Each year, these students are asked to develop a social action curriculum project – public service announcements, newsletters, unit plans, etc. – in collaboration with Elders in the community and teachers at the school and within the context of a curriculum design and evaluation course (see Ng-A-Fook 2011). TCs then implement the project at the school as part of their community service learning hours over the course of the academic year. As an element of the scope and sequencing of their professional development, the TCs travel three times to the community during the first semester, for an orientation and to plan their social action curriculum project. During the second semester, they travel once every 2 weeks to volunteer in a teacher's classroom. In May, after completing their final teacher education program, TCs have the opportunity to do a 7–10-day alternative placement at the school.

In the sections that follow, we illustrate the differing ways in which TCs experienced a gradual shift in their perspectives about the relational connections between local Indigenous cultural and environmental knowledges. We present our experiences with this program as a qualitative case study (Stake 2005), exploring the possibilities and limitations of using ecojustice as a model for engaging in EE toward reconciliation within the Canadian context of our Teacher Education Program at the University of Ottawa. Indigenous communities worldwide are extremely diverse, with their own unique cosmological, epistemological, ontological, and spiritual world views. Nonetheless, many of those differing world views are often united in terms of how their knowledge systems are rooted in relation to the land and waterways they live upon and across. As Marie Battiste (2013) observes: "Indigenous peoples have a science or way of knowing, but it is a concept that has embodied a way of life, an intimacy and directness with nature" (p. 160). Therefore, while our case study relates particularly to EE within the context of Canadian Indigenous communities, our initial research in this specific area of study provides some guidance for international teacher educators who seek to use EE to foster relations of reconciliation between Indigenous and non-Indigenous communities.

In the next section, we present a literature review which works to situate the concept of environmental reconciliation in relation to ecojustice education and Indigenous ways of knowing and relating to the world. In the third section, we present our qualitative interview data from six TCs completing a voluntary 7-day alternate practicum in an Indigenous community, demonstrating how their experiences transformed their perception of EE. In our final and concluding section, we provide a synthesis of the possibilities and limitations of our research in terms of recommendations for teacher educators, both within and beyond a Canadian context, who are committed toward developing EE within their teacher education programs as a potential praxis of environmental reconciliation in terms of our relations, not only with Indigenous communities but also with the more-than-human world.

2.1 Situating Ecojustice Education as a Praxis of Environmental Reconciliation

According to Rebecca A. Martusewicz et al. (2011), ecojustice education is a holistic, critical, and social justice-oriented approach to environmental education. One key element of ecojustice education is an emphasis on preserving the Commons, which can be described as:

> the necessary interdependent relationship of humans with the land, air, water, and other species with whom we share this planet, and the intergenerational practices and relationships among diverse groups of people that do not require the exchange of money as the primary motivation and generally result in mutual aid and support. (Martusewicz et al. 2011, p. 9)

Furthermore, these authors continue, ecojustice education "refuses the dichotomy between social justice and environmental concerns, arguing instead that they must be understood as grounded in the same cultural history" (p. 10). This recognition of the interrelation between the society and the environment represents an important step toward establishing EE as a model of environmental reconciliation.

Nonetheless, aspects of an ecojustice approach can be problematic from an Indigenous perspective. In her analysis of the work of ecojustice theorist Chet Bowers, Sandy Grande (2008) contends that Bowers maintains some of the very features he critiques in the Eurocentric tradition: "namely, the importance of critical reflection, an orientation toward (emancipatory) change, and a mastery of critical forms of literacy that enable such reflection and change" (p. 249). These features can be seen as products of the particular form of Western epistemology which philosopher Charles Taylor (1987) has described as characteristic of modern European cultures. This epistemology envisions detached and autonomous individuals operating within a universe of autonomous objects, stripped of all relationality. While Western cultures and epistemologies are no more monolithic than Indigenous ones, we believe it is important to recognize the broad patterns that characterize them, including the epistemological denial of environmental relationality and the impact such negation has had through the discursive regime of colonialism, namely, the institution settler North Americans call public schooling (Battiste 2013). Insofar as ecojustice remains rooted primarily in the Eurocentric philosophical tradition, therefore, it risks maintaining aspects of the status quo, despite its explicit critical intentions. Eve Tuck and Wayne Yang (2012), for instance, caution: "Claiming land for the Commons and asserting consensus as the rule of the Commons, erases existing, prior, and future Native land rights, decolonizing leadership, and forms of self–government" (p. 28). Martusewicz et al. (2011) acknowledge these concerns and dedicate a chapter of their book to describing parallels to ecojustice in Indigenous approaches as a way of "paying respect" (p. 250). This is an important gesture on their part. Nonetheless, a deeper engagement with Indigenous knowledges will enable ecojustice education, and EE more generally, to be more responsive to the cosmological, material, psychological, social, and spiritual needs and world views of Indigenous communities.

Following scholars such as Marie Battiste (2013) and Nicole Bell (2013), we understand Indigenous knowledges as being centered on places and relationships. Traditionally, Indigenous knowledges were grounded in a holistic awareness of what it meant to live well in a particular place. As Bell (2013) explains: "One's very existence depends on the web of interconnectedness between the self and the community and between the community and nature" (p. 98). As a result, a relationship to the specific local landscape was and is central to the ontology and epistemology of Indigenous communities (Simpson 2014). Kulnieks et al. (2010) have built on this understanding of knowledge as situated and interconnected to propose that all knowledges should be understood as originating from the contextual interpersonal relationships of particular communities. This perspective, however, stands in stark contrast to the Western ideals of knowledge that continue to dominate our educational institutions, which have tended to portray knowledge as fixed, absolute, and detached from any particular context (Bowers 2013). As a result, Battiste (2013) contends that education in Indigenous contexts must displace universalized forms of Western knowledge and return the relationality of local intergenerational knowledges toward decentralizing such essentialist settler colonial epistemological positions.

Dan Roronhiakewen Longboat et al. (2013) call for an ecojustice education that moves beyond dualisms and instead returns to a situated, holistic, sense of living well in the world. To achieve such (re)new(ed) balance requires, we suggest, carefully situated critiques of the universalizing ideologies put forth in the name of Western epistemologies (Taylor 1987) that have justified our historical route toward the current ecological crisis by portraying nature as an objective resource, existing apart from human life but available for human manipulation (Bowers 2013). As settler teachers and researchers, therefore, how might we then learn to become more self-critical in our past, present, and future engagements with Indigenous communities and their respective world views? How might we establish ethical relations with Indigenous knowledges without reducing their differing knowledge systems to ways that fit comfortably into our preexisting Eurocentric disciplinary teacher education categories (Tuck and Gaztambide-Fernández 2013)? Nonetheless, Longboat et al. (2013) also consider the separation of Western and Indigenous knowledges as another artificial dualism that must be overcome. Drawing on these scholars, therefore, we see the potential reconciling possibilities of reconceptualizing and renewing a relational partnership between models of ecojustice education developed in Anglo-Canadian educational institutions and traditional Indigenous ecological knowledge. Within our Canadian context, we have come to understand this partnership as a form of environmental reconciliation.

Through the lens of environmental reconciliation, we believe ecojustice education can provide a valuable curricular and pedagogical starting point in preparing TCs to teach EE in a culturally responsive manner in Indigenous communities. By disrupting the abstract and decontextualized knowledges of our educational institutions, and by calling attention instead to the concrete relational realities of the Commons, ecojustice education can unsettle many of the assumptions TCs bring with them into Indigenous communities, particularly the old (stock) Eurocentric humanist ideal that the environment is separate from society. However, ecojustice is not an Indigenous model and should be used with caution when preparing TCs for Indigenous contexts. In this light, we propose that Bell's (2013) portrayal of the Four Directions (Fig. 2.1) can act as a model of EE to complement and counterbalance an ecojustice model.

The primary value of Bell's model, for our situated purposes, is how it undercuts the type of analytic subdivision found in the traditional Eurocentric understanding of

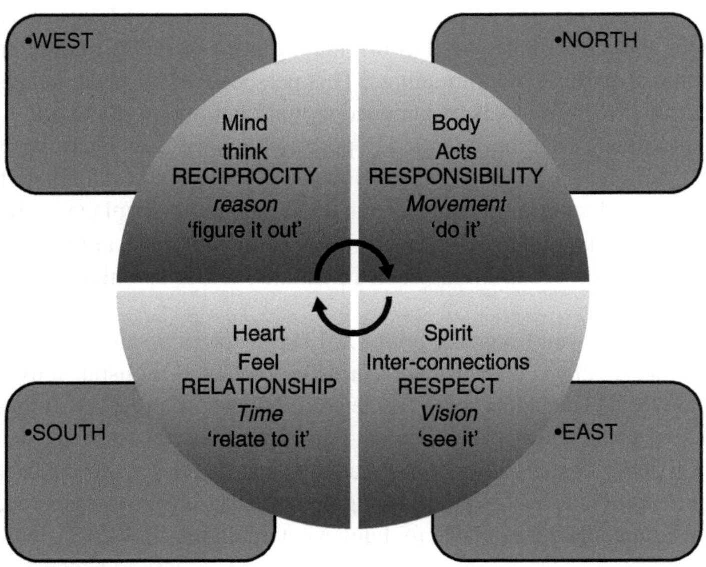

Fig. 2.1 The Four Directions (Source: Bell 2013 p. 96)

the self and the environment – what Bell refers to as "a breaking apart of concepts without relation to the whole" (p. 99). While neither self nor environment is explicitly included in Bell's model, a careful reading suggests that both are present throughout it but exist in a relational rather than an analytic form. A self is not an isolated, autonomous entity lurking in between mind, body, heart, and spirit but a dynamic result of their relational interactions. Likewise, the environment is not a reality separate from and existing beyond the boundaries of self but is dynamically related to mind, body, heart, and spirit – as self, society, and environment mutually constitute one another. Where the notion of environment in the Western tradition is often implied to be inert and objective, the Four Directions here are explicitly relational, for they are not merely a field within which the self can be placed, but are orientations a person always exists in relation to and lives by while seeking to live well.

Our adoption of Bell's model is meant to complement and contextualize ecojustice – not to replace it. Grande (2008) expresses concerns about the critical tradition in which ecojustice operates while ultimately observing the following:

> Nevertheless, if revolutionary critical pedagogy is able to sustain the same kind of penetrating analysis it unleashes on capitalism, it may evolve into an invaluable tool for indigenous peoples and their allies, fighting to protect and extend indigenous sovereignty over tribal land and resources. (p. 249)

We propose, therefore, that ecojustice education, when understood through Bell's model of the Four Directions, can help TCs learn how to live well within, and pedagogically reflect, the interconnected ecological and cultural contexts of the Indigenous communities where they serve. Such education must be a complex, iterative process. TCs should first be given the opportunity to explore general principles of Indigenous and environmental education in order to open critical engagement with their cultural assumptions and reflexivity about their teaching practices. As Cherubini (2011) suggests, TCs often do not have an understanding of cultures that differ from their own, and therefore teacher education programs should be challenging their ideologies and preconceptions, in order to develop a greater sense of self-awareness. This education should not remain at a general level, however, but should become situated as teachers immerse themselves into a particular Indigenous community and expose themselves to local traditional ecological knowledges. By revisiting and critiquing their cultural assumptions in an ongoing iterative and recursive manner, TCs can learn to take up EE not as a set of universal Eurocentric principles, but as a situated praxis of environmental reconciliation.

2.2 Nature's Kind of Like Their Best Friend

In this section, we present narrative snapshots of the critical reflections from six TCs who completed the voluntary 7-day alternate practicum in an Indigenous community. To that end, lead coresearchers Giuliano Reis and Nicholas Ng-A-Fook obtained research ethics to conduct the interviews with the teacher candidates after their practicum placements. The interviews lasted for about an hour each and were videotaped. Some of the questions included in the interview guide were:

"Has your understanding of the environment changed after practicum? If so, how?"; "Has your understanding of nature changed after practicum? If so, how?"; "Has your perspective on environmental education changed after practicum? If so, how?"; and "What benefits do you perceive there may be – if any – in integrating Indigenous knowledge, perspective and/or culture into your teaching practice?"

As we demonstrate, these TCs entered the community with a generalized awareness of common distinctive features of traditional Indigenous cultures, but they were not prepared to teach EE relationally, as a form of intercultural reconciliation. Some of these TCs were initially unsettled that the teachers were not, as they expected, explicitly incorporating references to EE into the curriculum. As we discussed in the previous section, this can be attributed to certain tendencies TCs learn within the discursive regime of a European intellectual tradition. Our findings suggest that these TCs entered their practicum expecting an analytical rather than a relational approach for incorporating EE into the classroom. Nonetheless, they appear to have gradually recognized that the environment was embedded coherently and holistically within the intergenerational teachings of this Indigenous community. We suggest that the learning process these TCs experienced over the course of their practicum, and in particular the preexisting assumptions they had to overcome, indicates the potential value of the situated and relational approach to EE that we have proposed.

At the outset, the TCs we interviewed appeared to enter the community assuming a binary division between the city and nature, expecting the Indigenous community to embody an urban idealization of naturalness. As Helen (pseudonyms are used throughout to preserve anonymity), one of the TCs, observed: "I'm not really a nature person. I'm a city girl which is going to be really interesting … on a reserve." In relation to EE, Helen expressed surprise that the resources of the local environment were not integrated into the curriculum. She spoke enthusiastically about the relaxed atmosphere near the lake and the comfort level of local community residents engaging in social and day-to-day conversations at the cottage where the TCs resided and coming together to have a meal. Helen concluded: "I'm just trying to think if there was anything really about the environment. No … not that I can think of." When asked whether environmental sustainability was integrated into the curriculum, Sarah, another TC, replied "No, not really." She did acknowledge the visible recycling program in place throughout the school: "They recycled. They had recycling everywhere. That's probably all for environmental." Shannon revealed similar first impressions. When asked whether EE was taken up in the school, she responded: "No not with the school as a whole." These comments reveal a particular set of preconceptions about what EE should look like. Coming from an urban context, these TCs seem to expect the environment to be taught separately, as a discrete topic of study. Furthermore, the praxis of EE was assumed to be limited to explicit conservation activities, such as recycling.

Nonetheless, through their experiences in the community, these TCs began to realize the complexity of the Commons. For instance, Shannon commented:

> They have Smart boards in all of their classrooms and I'm sure when they go home they have Internet as well but I felt like there must be some kind of an influence because when you leave the school and go home at night it's just … going to be your house and then the

> neighbours like way, way, way down that way and you know all through the night you can hear the lake, you can hear frogs, you can see stars. And I felt like it was overall a much calmer stress–free environment. And I noticed that in the kids as well.

The holistic and situated nature of the Commons in this Indigenous community led to EE being far more integrated and contextual than the TCs were originally expecting. Helen commented on one such instance:

> I guess the government wanted to cut down trees and they were trying to preserve their land and then they were talking later after we finished watching her video the students were talking about you know how they were losing their land, the loss of how important the land is to them … and then someone mentioned medicine 'cause … there's a lot of herbs and medicine they take directly from the land.

Abigail, meanwhile, commented on how EE was embedded within the very life of the community:

> They didn't give me a sense of protecting the earth–like there's a big focus on … protecting the environment … in Ontario curriculum and I didn't get that sense … from what they were talking about but … it's most like the nature's kind of like their best friend. You know what I mean, it's like a … part of their family. It's a crucial part of their society that they take along with them.

In the context of the school and its community, the environment was not being taught or experienced as a discrete subject, which exists apart from the curriculum and must be artificially integrated into it. Rather, the environment was implicit part of living well within community. As the TCs began to recognize this, it gradually changed their perspective on what it might mean to live and teach well within Indigenous communities.

As Bell (2013) explains, traditional Indigenous education was embedded within organic relationships between the person, the community, and the environment. As the TCs' narratives suggest, the day-to-day life of the school and its community was interconnected by deep intergenerational relationships with their history, culture, and environment. However, our analysis also suggests that the TCs were not well prepared to encounter this relational approach to EE in their practicum. In this sense, we propose that an ecojustice model of EE can be one resource to better prepare TCs for the complex interconnectedness of EE with Indigenous communities. While ecojustice maintains certain Western philosophical assumptions that limit its value as a possible praxis toward reconciliation of Indigenous and non-Indigenous communities, it can provide a valuable first step in disrupting the initial assumptions of TCs and preparing them to be open to the relationality of intergenerational Indigenous environmental knowledges. In our concluding section, we will provide some specific recommendations.

2.3 Teacher Candidates and EE Scholars Acting as Agents of Reconciliation

As we have suggested throughout this chapter, an ecojustice framework provides a valuable starting point for preparing TCs to recognize and engage with the complex relational interconnections of the environment within an Indigenous community.

Our findings indicate that an awareness of the traditional historical–cultural practices of different Indigenous communities is equally important for TCs, but that on its own, such awareness can simply become a new source of stereotypes. This generalized knowledge must be augmented with a relational awareness of how these traditions are preserved and transformed in the day-to-day reality and situated local contexts that remain specific to the different Indigenous communities who have inhabited these territories since time immemorial. We suggest, therefore, that the adoption of certain elements of ecojustice education within teacher education programs can help prepare TCs to encounter the interconnectedness of environmental and sociocultural education within an Indigenous community. By better preparing TCs to teach EE in a relational rather than a Western analytical and "enlightened" manner, we believe that these future teachers can be better equipped to act as agents of environmental reconciliation within and outside of Indigenous communities, both in Canada and internationally. In what follows, we recommend two specific principles of ecojustice education that can be used toward environmental reconciliation.

First, an ecojustice approach toward preparing TCs to engage with EE in Indigenous communities should begin with a critique of the Eurocentric logic of domination that was used to justify the establishment of settler colonial states like Canada. As Martusewicz et al. (2011) describe:

> We suggest that this logic of domination, deeply rooted in Western culture, and operating metaphorically, underlies the acceptance and continuation of class inequality, along with gender and race inequalities, other forms of social degradation, and ecological devastation. If the rational is superior, and humans are rational, then humans are morally justified in dominating or exploiting nature and anything else defined as analogous to or "like" nature. (p. 63)

This critical approach parallels Battiste's (2013) call for sustained critiques of the "cognitive imperialism" (p. 158) that continues to position Western knowledges as superior to Indigenous knowledges. In order for TCs to be able to engage respectfully with local Indigenous knowledges, they must first encounter the limits of their own epistemologies.

Second, an ecojustice approach emphasizes the importance of learning in collaboration with the community in relation to our local environments. Such situated learning is particularly important in relation to Indigenous education. As our small study indicates, even as TCs become more aware of Indigenous cultures, they will still tend to generalize this knowledge into abstract principles. This is a result, we suggest, of our Western educational system, which often valorizes abstracted knowledge (Bowers 2013). Our study also illustrated, however, how resilient these cultural assumptions are, even in the midst of situational learning. While our interviewees did gradually shift toward a more relational understanding of the interconnectedness of community, culture, and environment, such relational transformations and connections to the local environment could have happened more quickly if they had been better prepared within their teacher education program. This, in turn, would have enabled these TCs to make more effective use of their practicum, to the benefit of their Algonquin students and associate teachers. For this reason, we recommend having TCs visit an Indigenous school before their practicum. Particularly when bringing these experiences into dialogue with the critical approach to education previously discussed, TCs can reflect on these initial experiences as a way to work through their settler colonial assumptions before beginning their practicum.

Through official mechanisms such as Canada's Truth and Reconciliation Commission (TRC), settler colonial societies like Canada are becoming more aware of the need to bring reconciliation to our constitutional, community, and environmental relationships. As we have suggested in this chapter, one crucial area in which this work must take place is the preparation of TCs to teach in Indigenous communities. In particular, we have argued that they should be prepared to teach EE in a way that respectfully engages with the complex and situated interrelationship of culture and environment in specific Indigenous communities. We suggest that ecojustice education can provide a praxis to help prepare TCs to live within and pedagogically reflect the complex ecological and cultural contexts of the communities they serve. While ecojustice education has limits that must be acknowledged, we believe it can provide a valuable starting point, through which the initial settler colonial assumptions of TCs can be unsettled. In turn, TCs might become better prepared to take up EE within and in collaboration with Indigenous communities, whether that is here in Canada or abroad. Furthermore, these principles, drawn from Indigenous world views, can help international EE scholars and educators develop EE in a manner that is more responsive and related to situated and local contexts and, in turn, connected to other national and/or global environmental issues. In this way, both non-Indigenous TCs preparing to teach in Indigenous communities, and EE scholars working to make their work more responsive to Indigenous cultures, can take one small but important step toward being agents of reconciliation.

Questions

Discussion questions that could help instructors and students to engage in meaningful conversation about the ideas presented in this chapter.

1. What are ways in which Indigenous educational relations with the environment differ from those of non-Indigenous settlers in North America?
2. What are some strengths and weaknesses of ecojustice education, from the perspective of various Indigenous education scholars?
3. How might we draw on an Indigenous and ecojustice environmental education praxis to critique the cultural assumptions underlying existing curriculum policy documents and/or teaching practices both inside and outside the contexts of public education?
4. In what ways does our praxis in this course challenge and/or reproduce Eurocentric epistemologies that separate us from the relational context of the environment?
5. What are some ways we might engage a praxis of reconciliation with the environment as teachers and educational researchers?

References

Battiste, M. (2013). *Decolonizing education: Nourishing the learning spirit*. Saskatoon: Purich Publishing Ltd.

Bell, N. (2011). Creating shared understandings: Meeting indigenous education needs. In D. Stanley & K. Young (Eds.), *Contemporary studies in Canadian curriculum: Principles, portraits, & practices* (pp. 375–398). Calgary: Detselig Enterprises.

Bell, N. (2013). Anishinaabe Bimaadiziwin: Living spiritually with respect, relationship, reciprocity, and responsibility. In A. Kulnieks, D. R. Longboat, & K. Young (Eds.), *Contemporary studies in environmental and indigenous pedagogies: A curriculum of stories and place* (pp. 89–108). Rotterdam: Sense Publishers. https://doi.org/10.1007/978-94-6209-293-8_6.

Bowers, C. A. (2013). The role of environmental education in resisting the global forces undermining what remains of indigenous traditions of self sufficiency and mutual support. In A. Kulnieks, D. R. Longboat, & K. Young (Eds.), *Contemporary studies in environmental and indigenous pedagogies: A curriculum of stories and place* (pp. 225–240). Rotterdam: Sense Publishers. https://doi.org/10.1007/978-94-6209-293-8_12.

Butler, J. K. (2016). I live in a place that a river runs through: Localized literacy, currere, and a summer in an Ojibway community. *Journal of the Canadian Association for Curriculum Studies, 13*(2), 45–61.

Butler, J. K., Ng-A-Fook, N., Vaudrin-Charette, J., & McFadden, F. (2015). Living between truth and reconciliation: Responsibilities, colonial institutions, and settler scholars. *Transnational Curriculum Inquiry, 12*(2), 44–63.

Cherubini, L. (2011). Understanding the marginalized in the mainstream: Teacher education and aboriginal educational policy in Ontario (Canada). *International Journal of Education, 3*(2), 1–21. https://doi.org/10.5296/ije.v3i2.534.

Grande, S. (2008). Red pedagogy: The un–methodology. In N. K. Denzin, Y. S. Lincoln, & L. T. Smith (Eds.), *Handbook of critical and indigenous methodologies* (pp. 233–254). Thousand Oaks: SAGE. https://doi.org/10.4135/9781483385686.n12.

Henderson, J. (2000). The context of the state of nature. In M. Battiste (Ed.), *Reclaiming indigenous voice and vision* (pp. 11–38). Vancouver: UBC Press.

Henderson, J., & Wakeham, P. (2013). Introduction. In J. Henderson & P. Wakeham (Eds.), *Reconciling Canada: Critical perspectives on the culture of redress* (pp. 3–27). Toronto: University of Toronto Press.

Kulnieks, A., Longboat, D. R., & Young, K. (2010). Re–indigenizing curriculum: An eco–hermeneutic approach to learning. *AlterNative: An International Journal of Indigenous Peoples, 6*(1), 15–24.

Longboat, D. R., Kulnieks, A., & Young, K. (2013). Beyond dualism: Toward a transdisciplinary indigenous environmental studies model of environmental education curricula. In A. Kulnieks, D. Roronhiakewen Longboat, & K. A. Young (Eds.), *Contemporary studies in environmental and indigenous pedagogies: A curricula of stories and place* (pp. 9–20). Rotterdam: Sense Publishers. https://doi.org/10.1007/978-94-6209-293-8_2.

Martusewicz, R. A., Edmundson, J., & Lupinacci, J. (2011). *Ecojustice education: Toward diverse, democratic, and sustainable communities*. New York: Routledge. https://doi.org/10.4324/9780203836040.

McCrossan, M. (2015). Contaminating and collapsing indigenous space: Judicial narratives of Canadian territoriality. *Settler Colonial Studies, 5*(1), 20–39. https://doi.org/10.1080/2201473x.2014.925609.

Ng-A-Fook, N. (2011). Decolonizing narrative strands of our eco–civic responsibilities: Curriculum, social action, and indigenous communities. In K. Young & D. Stanley (Eds.), *Contemporary studies in Canadian curriculum: Principals, portraits, and practices* (pp. 313–341). Calgary: Detselig Enterprises Ltd..

Simpson, L. B. (2014). Land as pedagogy: Nishnaabeg intelligence and rebellious transformation. *Decolonization: Indigeneity, Education & Society, 3*(3), 1–25.

Stake, R. E. (2005). Qualitative case studies. In N. K. Denzin & Y. S. Lincoln (Eds.), *The SAGE handbook of qualitative research* (pp. 443–466). Thousand Oaks: SAGE.

Taylor, C. (1987). Overcoming epistemology. In K. Baynes, J. Bohman, & T. McCarthy (Eds.), *After philosophy: End or transformation?* (pp. 459–488). Cambridge, MA: MIT Press.

Truth and Reconciliation Commission of Canada. (2015). *Honouring the truth, reconciling for the future: Summary of the final report of the truth and reconciliation commission of Canada.* Retrieved from http://www.trc.ca/websites/trcinstitution/File/2015/Findings/Exec_Summary_2015_05_31_web_o.pdf

Tuck, E., & Gaztambide-Fernández, R. A. (2013). Curriculum, replacement, and settler futurity. *Journal of Curriculum Theorizing, 29*(1), 72–89.

Tuck, E., & Yang, K. W. (2012). Decolonization is not a metaphor. *Decolonization: Indigeneity, Education & Society, 1*(1), 1–40.

Veracini, L. (2010). *Settler colonialism: A theoretical overview.* Houndmills: Palgrave Macmillan. https://doi.org/10.1057/9780230299191.

Chapter 3
Reimagining Environmental Education as Artistic Practice

Hilary Inwood and Elizabeth Ashworth

Abstract This chapter explores the potential of art education to increase the power and reach of environmental learning in teacher education programmes and schools. The authors advocate for the inclusion of creativity, imagination and aesthetics as powerful pedagogical tools to provide a means for teacher educators of the arts to contribute towards positive environmental change. This entails a philosophical shift, one that better connects art education and environmental education, as well as a practical shift that reduces the waste and toxicity inherent to many art programmes. Descriptions of two different approaches to this reimagining of EE in Canada are provided. By sharing their experiences with integrating EE into their art education programmes, the authors discuss the benefits and challenges of reconceptualizing environmental education as artistic practice and aim to inspire others anywhere to do the same.

3.1 Reimagining Environmental Education as Artistic Practice

Creativity, imagination and aesthetics are not terms typically associated with the praxis of environmental education (EE). This is not surprising as over the past three decades, EE has found a strong foundation in science, geography and outdoor education (Palmer 1998), working towards a more sustainable future by developing learners' knowledge of human impact on the Earth's biological and physical systems. While some progress has been made in raising awareness about these environmental challenges (Klein 2015), we believe that large-scale change in terms of

H. Inwood (✉)
OISE, University of Toronto, 252 Bloor St. West, Toronto, ON M5S 1V6, Canada
e-mail: hilary.inwood@utoronto.ca

E. Ashworth
Schulich School of Education, Nipissing University, 100 College Dr,
North Bay P1B 8L7, ON, Canada
e-mail: liza@nipissingu.ca

G. Reis, J. Scott (eds.), *International Perspectives on the Theory and Practice of Environmental Education: A Reader*, Environmental Discourses in Science Education 3, https://doi.org/10.1007/978-3-319-67732-3_3

shifting patterns of living towards sustainability is still missing. How can EE be reimagined to help make sustainability a societal priority? What roles can creativity, imagination and aesthetics play in this? And how can teacher education help to facilitate this change? Examining the role that the arts can play in EE and in teacher education may help address these questions.

As faculty who teach in Canadian teacher education programmes, we are not alone in believing that initial teacher education should play a central role in ensuring that teachers have the knowledge, attitudes and skills needed to be environmental educators and advocates; our views are informed and supported by Charles Hopkins and Rosalyn McKeown (2005) and the Council of Ministers of Education, Canada (2012). As visual arts educators and practicing artists, we posit that the arts have an important role to play in raising societal awareness about environmental issues, as well as stimulating creative thought and innovative action in the movement towards sustainability. We support Joe Uehlein and Bren Smith's (2011) belief that "art can help us digest and make sense of what is happening in our world – a process essential for spurring political action" (n.p.). By sharing case studies in this area from the teacher education programmes in which we teach, we aim to bring together art making, research and teaching (referred to as a/r/tography by Irwin and de Cosson (2004)) as a means of starting a dialogue about the process of reimagining EE by exploring the roles that art education can play in EE within the context of initial teacher education.

3.2 Integrating EE and Art Education

> Here's the paradox: if the scientists are right, we're living through the biggest thing that's happened since human civilization emerged. One species, ours, has by itself in the course of a couple of generations managed to powerfully raise the temperature of an entire planet, to knock its most basic systems out of kilter. But oddly, though we know about it, we don't know about it. It hasn't registered in our gut; it isn't part of our culture. Where are the books? The poems? The plays? The goddamn operas? (McKibben 2005, online source)

McKibben is not the only one to call for a greater role for the arts in raising awareness about climate change and environmental degradation. While many visual artists (Edward Burtynsky, Noel Harding, Lynne Hull, Chris Jordan, Mierle Laderman Ukeles, Peter Menzel, to name only a few) have been making powerful contributions to environmental awareness and sustainability over the past 40 years, their work has been neither well-publicized nor widely understood by scholars in the EE field, nor by the general public. Visual arts scholars and critics such as John Beardsley (1998), Tim Collins (2007), Lucy Lippard (1998) and Barbra Matilsky (1992) have helped raise the profile of these artists' work, as have filmmakers such as Jennifer Baichwal (2006) and Thomas Riedelsheimer (2004). A variety of multidisciplinary programmes, such as the Cape Farewell project (http://www.capefarewell.com/), have also begun to recognize the need for scientists and artists to work collaboratively to make the interdisciplinary shifts towards sustainability.

Arts education has played only a peripheral role in EE in the past (Graham 2007). With a growing recognition that transmission-based, information-driven modes of learning are not sufficient in shifting attitudinal to behavioural change in relation to sustainability (McKibben 2005), the arts offer a more holistic approach, one that makes space for affective, creative and subjective approaches to learning. They provide an alternative means of developing the competencies of sustainable living by engaging the head (cognitive learning), the hands (embodied learning), the heart (affective learning) and the spirit, believed to be a route into authentic, long-lasting learning (Center for Ecoliteracy, "Discover" section). Certainly, as part of an inter-/trans-/multidisciplinary approach to EE, the arts offer the possibility of increasing the power and relevancy of environmental learning by appealing to a wider range of learners than ever before in creative and personal ways.

Artists, scholars and educators have been investigating artistic approaches to EE since the 1990s, sometimes referred to as environmental art education or eco-art education (Inwood 2009). Inspired by Suzi Gablik's (1991) articulation of a need for a social reconstructionist agenda in the discipline of the visual arts, many educators agreed with her call for "connective aesthetics" in art making (Gablik 1995, p. 84) and subsequently in art education. E. Louis Lankford (1997) framed this approach as "purposeful creativity" (p. 50), moving away from the art for art's sake model that was prevalent for most of the twentieth century and towards art making that had a purpose in addition to aesthetics. Art educators have begun to see the potential for a better integration of art education with the philosophical and theoretical trends of environmentalism, seen in the incorporation of place-based education and bioregionalism (Blandy and Hoffman 1993), ecofeminism (Hicks and King 1996) and systems thinking (Rosenthal 2003). These theoretical foundations continue to develop, with more recent contributions from Mark Graham's (2007) integration of a critical pedagogy of place with art education, Young Song's (2009) infusion of ecological art into EE and Tom Anderson and Suominen Guyas' (2012) articulation of the principles of E*art*h Education, built on the tenets of deep ecology.

Yet what is missing from the literature is an examination of how to bring art education and EE together in the context of initial teacher education and its potential for increasing the power and reach of EE. While some researchers have analysed the benefits and challenges of developing undergraduate courses in environmental art education (Inwood and Taylor 2012), there is a lack of understanding about its role in the context of teacher preparation. By describing and analysing case studies in this area at two different Canadian campuses, we aim to compare and analyse our approaches to reconceiving EE as artistic practice within the context of initial teacher education and argue for a more equitable inclusion of the visual arts in EE as a whole as a way to reimagine its praxis.

3.3 Turning Over a New Leaf at OISE

At the Ontario Institute for Studies in Education (OISE), in Toronto, Canada, I (Hilary Inwood) have taken a multipronged approach to integrating environmental art education into our initial teacher programme. Along with colleagues, we formed the Environmental and Sustainability Education (ESE) Working Group in 2008 in anticipation of the release of the Ontario Ministry of Education's (2009) EE policy framework, "Acting Today, Shaping Tomorrow". This policy was the first in Canada to formalize EE in the K–12 education system and proved to instigate some Ontario faculties of education to better embed EE in their preservice programmes. As Canada's largest teacher education programme (with over a thousand teacher candidates registered each year), we felt strongly that a clear presence for EE would support and deepen the creation of a culture of sustainability across the institution.

Each member of the group looked for ways to integrate EE within the existing content of our courses. While my challenge was timing (the art education module, mandatory for all elementary teacher candidates, was only 12 h in length), I realized that EE aligned nicely with the module's existing big ideas of curricular integration, growth and creativity. Over four classes, I highlighted the sustainability features of the classroom (a large blue bin for recycling, bins of reusable paper, found objects for art making and a vermicomposter for food waste). Wanting to remind my students of the pleasure of learning outside, I took the teacher candidates to a local park to draw in and from nature and to analyse a nearby public art installation focused on nature in the city, created by Susan Schelle and Mark Gomes. Another class was devoted to introducing environmental art education, contextualizing it in relation to the Ontario Ministry of Education's policy (2009), which advocates for an integrated approach to EE. Introduced to visual exemplars of what environmental art making looks like in elementary schools (created as part of my community-based artistic practice), teacher candidates became excited by the possibilities of this form of contemporary art practice and tried out similar ideas during their practicum placements.

By 2010, an elective course in EE was added to OISE's initial teacher education roster, providing a more formal venue in which to do this work. With teacher candidates immersed in a 36-h course devoted to EE, and an assignment that required them to pilot an EE learning experience, I found willing partners in the teacher candidates to explore how to incorporate artistic practices into EE in ways that could be shared across the institution.

During the first year, three teacher candidates offered to collaborate with me on an art project related to the Fatal Light Awareness Program (FLAP http://www.flap.org/). Dismayed by the discovery that migratory birds were hitting the glass windows of OISE's building, our team used artworks as a way to raise awareness about the plight of these birds in urban environments and to encourage OISE faculty to close their window blinds and turn off their office lights at night to deter birds from seeing the windows as fly-through portals. The project resulted in over 200 block relief prints (see Fig. 3.1) being delivered to offices throughout the institution, much

Fig. 3.1 Example of FLAP bird relief print

Fig. 3.2 Example of an energy conservation sticker

to the delight of faculty and staff. (At the time of the publication of this chapter, these prints were still found in offices throughout the building.)

Buoyed by this positive response and subsequent behavioural changes induced by the FLAP artworks, I promoted the idea of art projects as a means of raising awareness about environmental issues at OISE. In the second year, I ran three projects, two of which involved a partnership with Tara Rousseau, the art teacher at OISE's lab school, the Dr. Eric Jackman Institute of Child Study. I worked in collaboration with teacher candidate Hayley Chown and Tara's grade 5 class to create small drawings to encourage energy conservation, installed next to light switches in all OISE classrooms (see Fig. 3.2).

Hayley was so excited about this project that she enlisted other classmates to work on iterations of it in their teaching placements to support the Earth Hour event in March. I also worked with teacher candidate Stephanie Heim and Tara's grade 6 class to develop a clay-based art installation about the environmental rights of children. Even in grade 6, students could articulate the rights of children around the world to clean air, water and proper sanitation. As well, Aidan Hammond, a teacher candidate and experienced clay artist, worked alongside me on an installation that involved over a hundred teacher candidates visualizing the importance of nature-based learning in school and community settings. OISE students incised and

Fig. 3.3 Clay installation on nature-based learning

sculpted their impressions and memories of nature-based learning into clay relief tiles, helping to remind viewers about the importance of reconnecting children and adults with nature (see Fig. 3.3).

These projects were the starting point of OISE's "Take the Stairs" campaign. Installed in OISE's main stairwell as a walking art gallery, these artworks aimed to encourage the OISE community to take the stairs, rather than using the elevators, thereby conserving energy and improving the health and wellness of walkers. The projects were a success, garnering attention both within and outside the building and spurring the development of three more eco-art installations the following year. Grade 1 learners from the lab school created two tall murals to share their knowledge-building approach to learning about trees, which involved learning through observation, discussion and research, collaboratively building their understanding over time. Working in collaboration with Tara, me and OISE teacher candidate Jennifer Sharpe, the class created two large-scale paintings to share what they had learned, showing two trees in different states of seasonal change, with their text-based descriptions forming part of the colourful images (see Fig. 3.4).

Since the inception of this unusual art gallery, we have added a number of other artworks. Grade 3 students from the lab school shared their study of First Nations peoples, the cardinal directions and native plant species by creating artworks using paint, clay, photographic transfers and poetry. In 2013, teacher candidates helped create an art installation focused on sustainability actions called "Turning over a New Leaf". It was made up of dozens of clay leaves inscribed with images and

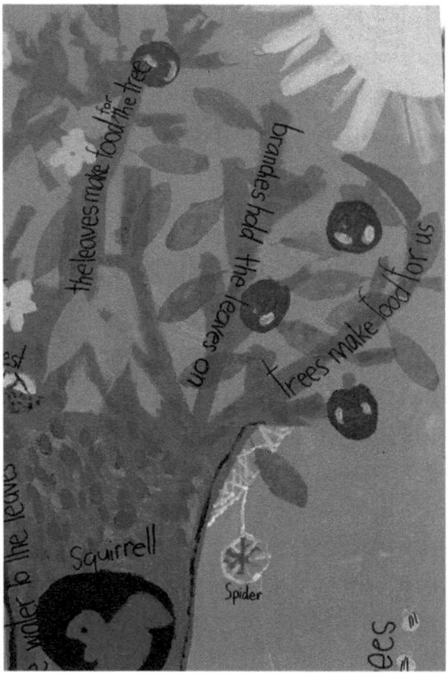

Fig. 3.4 Detail of grade 1 mural of trees

messages that remind viewers of the actions they can take to live more lightly on the Earth. Another iteration of the FLAP project has also been installed, involving hundreds of watercolours of local birds in flight, as well as an installation focused on the plight of Ontario bees. Teacher candidates will use the time they put into these projects towards fulfilling the requirements of OISE's Environmental Leadership Certificate, a means of recognizing their commitment to 36 h of formal, informal and service-based learning related to EE.

What lessons have been learned from this ongoing experimentation into environmental art education at OISE? Teacher education as a whole needs to be reimagined to align with beliefs and practices about environmental sustainability (Hopkins and McKeown 2005), and modelling what sustainability can look like in art education classrooms is a vital part of this. Integrating EE into all subject areas in an initial teacher education curriculum is important, as teacher candidates, much like younger students, enjoy learning experientially, holistically and affectively. Another lesson learned is that one of the tenets of EE, start small and do projects well, is just as applicable in eco-art education. Showing that artworks and installations made in environmentally friendly ways (using natural materials such as clay) can be completed on time and on budget is crucial for ensuring ongoing support. Art installations, when done collaboratively, can be created quickly and inexpensively; many hands make for light work. We also learned that allies in this process can be found in unexpected places; for example, in my experience, facilities managers are often happy to have more art in buildings, especially when the art is made by those who use the building. With a little forethought and imagination, any art installation can be focused on socially and politically relevant big ideas, helping learners to manifest their theoretical learning in concrete and aesthetic ways. For those who often don't see themselves as creative, teacher candidates are thrilled to become artistic collaborators on projects that leave a lasting mark, thus modelling encouragement and support for them demonstrates, in an explicit way, what they can do as teachers in their own future classrooms.

3.4 Sustainable Studio Thinking at Nipissing University

For many years, I (Elizabeth Ashworth) have been aware of how Hilary infused her passion for EE into her art, research and teaching at OISE (Inwood 2010), and she inspired me to experiment with similar practices in my art education courses at Nipissing University, in North Bay, Canada. I work with teacher candidates to create a variety of drawings, paintings, prints and sculptures as exemplars for their future classrooms. During the process of creating their art education sample portfolios, they use an abundance of supplies, mimicking traditional visual arts programmes that rely heavily on consumable supplies. Once the artworks are assessed, however, these are often discarded by the teacher candidates, resulting in large amounts of paper and other materials being used on a short-term basis. As the recycling programme within the City of North Bay does not have a wide breadth of acceptable items, the potential for things to go into landfill is high. Like Pamela Taylor (1997), I reached a tipping point when I witnessed teacher candidates thoughtlessly discarding materials in this process.

In an effort to stop these wasteful practices, I asked teacher candidates to consider their roles as art education leaders and encouraged them to reimagine art programmes through an EE lens. I embarked on a journey to plan, implement and assess a sustainable studio. Among others, I considered Lorraine Chiarotto's (2011) ideas for incorporating environmental expectations into school projects and Mark Stewart's (2010) practical concepts for encouraging sustainability in higher education. Except for Pamela Taylor (1997), no one included strategies on how to reduce art programme refuse in educational contexts.

I then re-evaluated the art education classroom at Nipissing and created short- and long-term plans for transforming it into a sustainable teaching and learning space and identified areas in our programme where sustainable practices could be improved. Each year, for example, we ordered dozens of Plasticine bricks for teacher candidates to create claymation figures. At the end of each term, however, only some were taken home; leftover works were tossed into boxes for recycling. In order to encourage both sustainability and creativity in the sculpture unit, I asked the teacher candidates to imagine moving into a classroom where they found boxes of used Plasticine. They could neither buy new materials nor discard the used supply, so they had to consider ways to use it creatively. They appreciated the real life scenario, and once they began building their sculptures, they started to accept the blended colours. As a result, fewer works were left behind, and those that were discarded were donated to a local school for use in their claymation projects (see Fig. 3.5).

Fig. 3.5 Sculptures created with recycled Plasticine

Next, I embedded the concept of sustainability into all lessons and assignments. For instance, at the beginning of each term, I took the teacher candidates on a tour of the art classroom, placing emphasis on the use of the recycle bin, hand dryers, cloth towels and the absence of garbage cans and paper towels; I wanted them to think critically about classroom waste from the outset. I then gave them the choice to create sketchbooks with faded paper or use free sketchbook apps (Autodesk Sketchbook, Draw Color, Drawing Desk) on their tablets.

After this introduction, I modelled and reviewed sustainability considerations in all lessons; for example, teacher candidates made sculptures out of recycling bin materials and created paintings with used items (plastic lids for palettes; old sponges and toothbrushes as paint brushes). They participated in a group project where they integrated visual arts with another subject (Steele and Ashworth 2013) or an educational focus (such as aboriginal perspectives, literacy or social justice). I added the requirement that the art component must be made of recycled materials and found these works far more interesting as it encouraged the teacher candidates to think more critically and creatively when planning and constructing their works (see Fig. 3.6).

I also worked with a colleague to take advantage of our university's location, adjacent to a natural area. We led extracurricular workshops where teacher candidates created teaching exemplars out of natural items found within the forest. Without access to traditional art supplies, most groups created works inspired by artist Andy Goldsworthy (see Fig. 3.7). All works were made and discussed in situ, then photographed and left for nature to biodegrade.

After 1 year of implementing these various sustainable practices, I assessed their strengths and weaknesses to make constructive changes for future courses; for example, I kept track of the use/nonuse of consumable supplies and documented teacher candidates' use of recyclable and natural materials in their artworks. Although the consumable waste had reduced dramatically, I found that teacher candidates were still discarding artworks they did not want. In order to address this

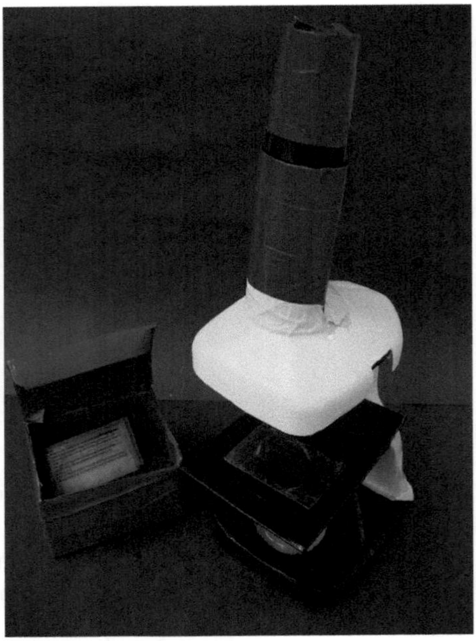

Fig. 3.6 Working microscope created from a detergent bottle

Fig. 3.7 Bush Lion: created from items found on the ground

waste, I encouraged them to keep all of their works. Now one of their portfolio requirements is to create at least one work of art using bits and pieces of works they do not want to keep (see Fig. 3.8).

Recently, another opportunity for integrating EE and art education at Nipissing arose: the addition of two new elective courses, "Environmental Education Across the Curriculum" and "Teaching, Learning, and Being in the Outdoors". Both courses have the potential to incorporate art, research and teaching with sustainable practice considerations.

Fig. 3.8 Storm: a mixed media piece created from discarded artworks

Overall, my intention has been for teacher candidates to become eco-literate art educators. I want them to experience and reflect on sustainable studio practices and then take what they learn about the environmental, economic, community and cultural benefits of teaching and learning in a sustainable studio/classroom into their future schools. They can easily integrate their environmental art education knowledge with other subject areas and share these with classroom teachers who host our teacher candidates during their practice teaching placements. By taking their EE knowledge beyond the university, and sharing sustainable studio practices learned in their teacher education programme, it is my hope that teacher candidates will inspire other educators to create similar environments in their own classrooms.

3.5 Connections and Insights

Both of our cases reveal the benefits and challenges of reimagining EE through artistic practice in different ways. Moreover, showing our teacher candidates that the praxis of EE can be integrated into art education was easier than we initially imagined, and we demonstrated that art has a variety of social, cultural and political purposes beyond the aesthetic, modelling Gablik's (1991) social reconstructivist approach. By connecting art education to environmental sustainability in practical and theoretical ways, we made EE an explicit and integral part of our teaching. We have witnessed the personal transformation of our students, many moving from an "I'm not an artist" mindset to reimagining themselves in a new light, as both creative people and advocates of sustainability. Seeing themselves aligned with contemporary artists who are using sustainable approaches to art making has been empowering for many and has made us believe that our teacher candidates are more likely to implement this approach in their own classrooms in practicum and beyond.

This sustainable approach to creativity has benefited our universities in a variety of ways: from reducing the amount we spend on studio supplies to decreasing the amount of garbage our students create. Our eco-art endeavours have improved the aesthetics of some of our institutional spaces and raised awareness about a wide range of solutions to environmental issues, including waste minimization, energy conservation and habitat preservation. Finally, we have influenced some of our colleagues to take sustainability more seriously and to reconsider the ways in which it can be integrated into all areas of teacher education.

Although we have had many successes, we have also faced challenges and limitations in this work. It has been difficult to get our students (and colleagues) to change their habits around the use of disposable coffee cups, water bottles and paper towels; not everyone in our workplaces embraces our passion for sustainability. Similarly, some colleagues are not open to integrating EE into their teacher education courses; from their perspective, it is yet another addition to an already overloaded curriculum, and they lack the knowledge and awareness (or interest) into how to infuse EE into their course. Some of our managers are very pleased that our use of recycled and natural materials has resulted in a reduction in the use of consumable materials and have reduced budgets with this in mind. Certainly, we have not gotten fully to where we would like to be in terms of a zero waste approach; artistic practice and school-based learning involves the use of consumable supplies that generate waste, so it is still a dream to imagine a fully zero waste art education programme.

This process has also left us with many questions moving forward. How can we grow our efforts beyond our subject area and get more faculty members on board with this reimagining of teacher education? How can we get support in this work from the companies that make art supplies? And what impact is this work having on our students' learning? Are our students taking the practices they have learned into their practicum schools, and are they able to successfully implement the practices we have shared with them with their own students? Are they able to translate this new knowledge into their personal lives and live in more sustainable ways? Anecdotally, we are hearing many stories from our present and former students about shifts in their attitudes, as well as their personal and professional behaviours, towards living more lightly on the earth. But there is no doubt that a more formal study is needed in this area to better understand the impact of the changes we have made in our teaching practice and if they are resulting in long-term changes in our students' lives.

We recognize that the impact of our efforts, so far, is local, but has the potential to reach far beyond our classrooms. We have created goals with which to work: first, to integrate EE more fully into our art education courses, our universities and our communities and, second, to model how to integrate creativity, imagination and aesthetic practice into positive sustainable changes for our teacher candidates and colleagues. The new 2-year teacher education programme in Ontario offers the possibility to develop our ideas with students over a longer period of time. We see the efforts described in this chapter as first steps towards a greater reimagining of art education and teacher education to support EE and welcome others to join us in this journey.

Questions

Discussion questions that could help instructors and students to engage in meaningful conversation about the ideas presented in this chapter.

1. Both Inwood and Ashworth faced challenges in their planning and implementation of sustainable practices in their teacher education classrooms. What challenges could you face in your workplace regarding a move to more environmental awareness and sustainable practices?
2. The study of eco-art/artists, and the practice of making art with recycled and natural materials, "transformed" their teacher candidates. How might the same happen with your learners?
3. Inwood and Ashworth focused on integrating visual arts and environmental education within their initial teacher education programmes. How could this integration be imagined and implemented in other curricular subjects and areas of focus in elementary, secondary and post-secondary institutions?
4. The authors advocate for the inclusion of creativity, imagination and aesthetics as powerful pedagogical tools for teachers to contribute towards positive environmental change. Consider how a lesson (or unit) you have taught could be reimagined in this context
5. Modelling positive sustainable practices in education is a key theme in this chapter. What steps could you take in your workplace and community to be a leader in environmental education?

References

Anderson, T., & Guyas, A. S. (2012). E"art"h education, interbeing, and deep ecology. *Studies in Art Education, 53*(3), 223–245.

Baichwal, J. (2006). *Manufactured landscapes [documentary]*. Toronto: Foundry Films & the National Film Board of Canada.

Beardsley, J. (1998). *Earthworks and beyond: Contemporary art in the landscape*. New York: Abbeville Press.

Blandy, D., & Hoffman, E. (1993). Toward an art education of place. *Studies in Art Education, 35*(1), 22–33. https://doi.org/10.2307/1320835.

Center for Ecoliteracy. (n.d.) Competencies for sustainable living. Retrieved from http://www.ecoliteracy.org/taxonomy/term/84.

Chiarotto, L. (2011). *Natural curiosity: Building children's understanding of the world through environmental inquiry*. Toronto: University of Toronto.

Collins, T. (2007). *Lyrical expression, critical engagement, transformative action: An introduction to art and the environment*. Retrieved from http://wayback.archive-it.org/2077/20100906195312/, http://www.communityarts.net/readingroom/archivefiles/2003/06/lyrical_express.php

Council of Ministers of Education, Canada. (2012). *Education for sustainable development in Canadian faculties of education*. Retrieved from http://www.cmec.ca/Publications/Lists/Publications/Attachments/279/ESD_Dean_reportEN.pdf

Gablik, S. (1991). *The re–enchantment of art*. New York: Thames and Hudson.

Gablik, S. (1995). Connective aesthetics. In S. Lacy (Ed.), *Mapping the terrain: New genre public art* (pp. 74–87). Seattle: Bay Press.

Graham, M. (2007). Art, ecology, and art education: Locating art education in a critical place–based pedagogy. *Studies in Art Education, 48*(4), 375–391.
Hicks, L., & King, R. (1996). Ecofeminism, care and the environment: Towards a greening of art education. In G. Collins & R. Sandell (Eds.), *Gender issues in art education: Content, contexts and strategies* (pp. 90–101). Reston: National Art Education Association.
Hopkins, C., & McKeown, R. (2005). *Guidelines and recommendations for reorienting teacher education to address sustainability*. UNESCO Education Sector. Retrieved from http://unesdoc.unesco.org/images/0014/001433/143370e.pdf
Inwood, H. (2009). *Artistic approaches to environmental education: Developing eco–art education in elementary classrooms*. Doctoral dissertation. Retrieved from ProQuest Dissertations & Theses Global. (NR67360).
Inwood, H. (2010). Shades of green: Growing environmentalism and sustainability in art education. *Art Education, 63*(6), 33–38.
Inwood, H., & Taylor, R. W. (2012). Creative approaches to environmental learning: Two perspectives on teaching environmental art education. *International Electronic Journal of Environmental Education, 2*(1), 65–75.
Irwin, R., & de Cosson, A. (Eds.). (2004). *A/r/tography: Rendering self through arts–based living inquiry*. Vancouver: Pacific Educational Press.
Klein, N. (2015). *This changes everything*. Toronto: Knopf Canada.
Lankford, E. L. (1997). Ecological stewardship in art education. *Art Education, 50*(6), 47–53. https://doi.org/10.2307/3193688.
Lippard, L. (1998). *The lure of the local: Senses of place in a multicentered society*. New York: The New Press.
Matilsky, B. (1992). *Fragile ecologies: Contemporary artists' interpretations and solutions*. New York: Rizzoli.
McKibben, B. (2005). *What the world needs now, is art, sweet art*. Retrieved from http://www/grist.org/comments/soapbox/2005/04/21/mckibben-imagine/index.html
Ontario Ministry of Education. (2009). *Acting today, shaping tomorrow*. Toronto: Queen's Printer for Ontario. Retrieved from http://www.edu.gov.on.ca/curriculumcouncil/shapetomorrow.pdf.
Palmer, J. (1998). *Environmental education in the 21st century: Theory, practice, progress and promise*. London: Routledge.
Riedelsheimer, T. (2004). *Andy Goldsworthy: Rivers and tides, working with time [Documentary]*. New York: New Video Group.
Rosenthal, A. (2003). Teaching systems theory and practice through environmental art. *Ethics and the Environment, 8*(1), 153–168. https://doi.org/10.1353/een.2003.0013.
Song, Y. I. K. (2009). Community participatory ecological art and education. *International Journal of Art & Design Education, 28*(1), 4–13. https://doi.org/10.1111/j.1476-8070.2009.01588.x.
Steele, A., Ashworth, E. (2013). Walking the integration talk: An artsci project. *Canadian Journal for the Scholarship of Teaching and Learning*, *4*(2). doi:https://doi.org/10.5206/cjsotl–rcacea.2013.2.6.
Stewart, M. (2010). Transforming higher education: A practical plan for integrating sustainability education into the student experience. *Journal of Sustainability Education, 1*, 195–207. Retrieved from http://www.jsedimensions.org/wordpress/2010-inaugural-issue/.
Taylor, P. (1997). It all started with the trash: Taking steps toward sustainable art education. *Art Education, 50*(2), 13–18. https://doi.org/10.2307/3193638.
Uehlein, J., & Smith, B. (2011). *Calling all artists: The climate change movement needs you!* Retrieved from-http://grist.org/climate-change/2011-05-13-calling-all-artists-the-climate-movement-needs0you/

Chapter 4
Integration, Inquiry and Interpretation: A Learning Garden Alternative Placement and Eco-mentorship Program for Pre-service Teachers

Kelly Young and Darren Stanley

All education is environmental education.

Orr 19 92, p. 81

Abstract In this chapter, we address the complex nature of environmental education by highlighting some of the ways in which language—specifically, root metaphors—perpetuates the ongoing disconnection of humans and the natural world. We make connections to insights about an established learning garden alternative placement and eco-mentorship program that reflect sustainable environmental education (EE) practices and help pre-service teachers re-establish a deeper connection between themselves and their natural environments.

To address the complexities of EE whereby all things are interrelated and interconnected, scholars have suggested that the educational system must consider more than scientific academic disciplines. For instance, addressing language that perpetuates a disconnection of humans and the natural world as a response to the complex nature of EE, we have envisioned particular curricular practices as part of a learning garden alternative placement and eco-mentorship program (Martusewicz, Edmundson and Lupinacci). This placement and program, embedded in Trent

K. Young (✉)
School of Education, Trent University, 1600 West Bank Drive,
Peterborough, ON, K9L 0G2, Canada
e-mail: kellyyoung@trentu.ca

D. Stanley
Faculty of Education, University of Windsor, 401 Sunset Avenue,
Windsor N9B 3P4, ON, Canada
e-mail: dstanley@uwindsor.ca

G. Reis, J. Scott (eds.), *International Perspectives on the Theory and Practice of Environmental Education: A Reader*, Environmental Discourses in Science Education 3, https://doi.org/10.1007/978-3-319-67732-3_4

University's School of Education program in Peterborough, Ontario, for Teacher Candidates (TCs), reconceptualizes EE as an integrated endeavour that seeks to address the ongoing hyper-separation of humans and the natural world. We start this chapter with a view of some of the ways in which language can be problematic for understanding environmental education.

4.1 How Language Contributes to a Crisis of Disconnection and Alienation

As part of a larger research study, we have been engaging TCs in curriculum methods for ecological literacy development while considering various issues pertaining to human relations with the natural world. In the context of a learning garden alternative placement and eco-mentorship program, we have been working to identify methods that foster an analysis of the ways in which language obfuscates relationships amongst humans and the natural world. Thus, our method involves an analysis of language to nurture the development of ecological literacy in TCs, addressing the linguistic roots of the ecological crisis. In doing so, we draw upon an eco-justice framework, defined as:

> The condition or principle of being just or equitable with respect to ecological sustainability and protection of the environment, as well as social and economic issues. (OED 2008)

Put differently, for us, eco-justice education involves a linguistic analysis of the ecological crisis, utilizing a framework developed by Chet Bowers (2004), to address the invisible, damaged or otherwise lost connections between humans and the natural world. A linguistic analysis of the ecological crisis of disconnection and alienation is complex and requires us to attend to the entanglement of worldly relationships. Bowers (2003) believes that to understand an eco-justice framework, one has to begin with language, particularly root metaphors that frame our social and environmental relationships in rather mechanistic ways. Bowers (2004) offers the following view of mechanistic root metaphors:

> Languaging processes carry forward past ways of thinking that are based on assumptions unique to the culture; these deeply held and generally taken-for-granted assumptions, which are derived from the culture's mythopoetic narratives and powerful evocative experiences, are encoded in the words that are called root metaphors; the root metaphors of a culture provide the interpretative frameworks that survive over many generations and influence values, approaches to problem solving and activities in a wide range of daily life; the root metaphors, as meta-cognitive schemata, also influence the silences as well as what will be marginalized; the dominant root metaphors in the West that have contributed to an ecologically destructive culture include mechanism, a linear interpretation of progress, anthropocentrism, Cartesian individualism, patriarchy, and, now, evolution as a way of explaining which cultures will survive; these root metaphors reproduce the pre-ecological ways of thinking, and are also basic to the continued expansion of the industrial culture…. (n.p.)

Thus, through an analysis of the history and evolution of language, we can see how our attitudes, beliefs and, ultimately, Western culture came to exist. Further, Bowers

(2004) also moves beyond mechanistic root metaphors to include ecological root metaphors in his interpretive framework, as he writes:

> The root metaphor that serves as an interpretive framework for addressing eco-justice issues is ecology-which highlights awareness of relationships and interdependencies with the commons; as a root metaphor, ecology locates the individual as a participant within the ecological systems that we are calling the commons, which is profoundly different from how an anthropocentric root metaphor (interpretative framework) leads to thinking of oneself as an observer, a person who appropriates the environment for personal gain, or as totally indifferent to the changes occurring in the environment. (n.p.)

Highlighting the importance of recognizing both mechanistic and ecological root metaphors is part of the work of eco-justice education and follows David Orr's (1992) suggestion that "all education is environmental education" (p. 81). For over a decade, we have drawn upon Bowers' (2004) framework to address language issues as we teach eco-justice education at our respective institutions. Further, we have proposed that an eco-justice framework be used in a learning garden alternative placement and eco-mentorship program to address the conventional view that EE is strictly part of a science curriculum (Longboat et al. 2013). Put differently, from an eco-justice perspective, we have been working on a program for TCs to reconceptualize environmental curriculum in terms of a linguistic analysis that can inform all curricula. In the coming sections, we briefly describe how we have engaged in language analysis as part of our curriculum, documenting these eco-justice methods. We also share our insights about the TCs' perceptions about the eco-mentor programming and their understanding of ecological literacy.

4.2 Pre-service Learning Garden and Eco-mentorship Programming

In response to a need for infusing environmental curriculum with a linguistic analysis, we have been involved in two distinct programs for ecological literacy development. That is, since 2011, Trent University's School of Education in Peterborough, Ontario, has worked with a number of partnerships with organizations from the local community to serve as sites of ecological inquiry for TCs. We briefly describe, here, two of those partnership initiatives: a learning garden opportunity and an eco-mentorship program.

In 2011, the School of Education partnered with Camp Kawartha, an established outdoor education camp located in the Kawartha Lakes of Ontario, Canada. Numerous cohorts of TCs participated in four 3-h workshops as part of the school's eco-mentorship program. Focusing on EE themes, students drew upon local natural settings to teach about an ethic of care for the environment, learn how language obfuscates different environmental concepts and ideas, experience greater hope about EE, construct cross-curricular strategies for EE and infuse Indigenous knowledge across the curriculum (Bell et al. 2013).

In the following year, the School of Education and a local well-established not-for-profit Ecological Park/GreenUP partnered to offer environmental curricula to TCs. The curricula promotes ecological literacy and food sustainability through the use of community gardens as places for meaningful, active, integrated learning. This learning garden alternative placement involved TCs in the planning of gardens; greening local schools through projects like the creation of compost boxes, bird feeders and rainwater barrels; and attending workshops. Put differently, the learning garden field placement involved building capacity for TCs to develop, implement and maintain learning garden programs in local schools. For example, TCs work in a local park with an expert gardener to learn essential skills about planning and planting gardens, pollinators and composting. Subsequently, they were provided with opportunities in local schools to plan and plant school-based gardens. Concurrently, they taught about food sustainability and the importance of locally grown vegetables to K-12 students through integrated curricula and provided environmental leadership to students.

As researchers, we have collected data on the learning garden and eco-mentorship programs since their inception. Specifically, we have studied how cohorts of the TCs have participated in a 75-h alternative placement involving workshops on garden planning, food sustainability and strategy development for integrating EE across the curriculum. We have also studied how TCs have participated in four 3-h workshops (as described above) focused on EE and developed action projects that were then implemented in their practicum placements in K-12 classrooms in the Peterborough area. This research initiative was in response to the policy framework for environmental education in Ontario schools (Ontario Ministry of Education 2009).

4.3 Root Metaphors and the Language of an Eco-justice Curricula

In this section, we reflect upon a particular workshop that addresses language and the critical role it plays in our disconnection and alienation to EE to frame our emerging views of the importance of eco-justice-oriented EE curricula. Specifically, we outline the various ways in which eco-justice addresses the interrelationality of all living things by paying particular attention to the theory-to-practice approach that brings together the work that the students do. More importantly, we consider the ways in which an eco-justice framework has helped these students address and alleviate the disconnection and alienation to EE by illuminating root metaphors that perpetuate anti-ecological habits of mind.

Our curriculum contributions to the Learning Garden and eco-mentorship programs involve a workshop for the TCs on developing an eco-justice framework in K-12 classrooms. We find that language involves ideas, systems of metaphors and stories that shape epistemologies. As such, during our workshop, we introduce the concept of root metaphors. Western, or Eurocentric, ways of knowing involve anthropocentric framing, conceptualizations of humans beings as machines that

Table 4.1 An eco-justice linguistic hierarchy of binaries chart

Thinking via binaries and mechanistic metaphors (like a machine)		
Human-constructed categories mediated by language		
Mechanism (eurocentric colonizing discourses)	Binaries	
Anthropocentric	Human	Nature
	Culture	Nature
Andropocentric	Man	Woman
Rationalism/scientism/ableism	Mind	Body
	Reason	Emotion
Ethnocentricism/racism	Civilized	Savage
Individualism	Self	Community

Adapted from Martusewicz et al. (2011)

process information and hierarchal dualisms that have organized human language and discourse into knowledge (Martusewicz et al. 2011).

As described earlier in this chapter, root metaphors are languaging processes that influence the ways in which humans interact with the natural world. We have found that mapping linguistic aspects helps us to name key hierarchical binaries that structure ways of thinking, knowing and being in the world. Such a map reveals various binaries and mechanistic root metaphors that are human-constructed categories mediated by language. By naming the mechanistic metaphors at work that also represent key hierarchical binaries in our daily language, it is easier to see how human constructed categories continue to perpetuate inequalities. Further, these root metaphors are derived from mythopoetic narratives that influence ways of thinking and being (Bowers 2003). The following chart illustrates an eco-justice linguistic hierarchy of binaries (Table 4.1).

An exercise in the mapping of these binaries, in fact, may help learners to notice how one set of these binaries almost entirely alludes to typically female traits or nature-as-woman-as-emotions, whereas the other binaries suggest typically male traits or man-as-mind-as-reason. Put differently, the male traits represent dominant human-constructed categories, and the female traits represent subordinate categories. As such, humans dominate nature, men dominate women and so on. By studying this chart of binaries, learners can see the interconnectedness of all oppression, that is, oppression of the natural world, anthropocentrism; oppression of women, andropocentrism; oppression of the mind-body relationship; oppression of ethnicities or people of colour, ethnocentrism; and a privileging of individuals over community relationships. Although this chart is helpful in providing an historical overview, it is often difficult, however, for most teachers to apply this to daily life in the classroom as these binaries are taken for granted in our languaging processes.

In the classroom and the field, we have found that it is easier to help TCs see the ways in which language can illuminate or hide taken-for-granted associations through a review of word associations. Beginning with the manner in which the universe is likened to a machine, we turn to Bowers (2003) who writes that a "mechanism is a root metaphor that had its roots in the transition from a Medieval to the

Table 4.2 Root metaphors that connect and disconnect us from the natural world

Forest	Timber
Trees	Lumber
Insects	Pests
Soil	Dirt
Plants	Weeds
Animals	Critters/livestock
Landfill	Garbage
Rocks	Aggregates

modern worldview" (p. 2). Furthermore, Bowers also points out that Johannes Kepler "used the root metaphor of mechanism to explain the universe as a celestial machine" (p. 2). Just as the universe was likened to a machine, so, too, was the body and, of course, natural and nonnatural elements of our world. For example, words like stream and drain bring attention to the mechanistic nature of the latter (Martusewicz et al. 2011). In addition, we have often found that during workshops, when asked to brainstorm about the associations that words like stream and drain conjure up, stream is linked to something ecological and natural, whereas drain is linked to something that is human made and something that one can empty chemical products into when there is a clog.

Ultimately, we discuss with students how a stream and drain are alike in that they lead to the same natural ecosystem. What is interesting, however, is how the word drain seems so nonnatural and mechanistic that they easily dissociate it with its role in our ecological surroundings. This is one example of how we need to name taken-for-granted associations that can lead us either away from or closer to deeper connections between language, culture and natural environments. Consider the following word associations in Table 4.2:

By engaging students in a linguistic and cultural analysis of ways in which root metaphors may suggest disconnections between humans and the natural world, they can begin to understand how the environmental crisis is a cultural crisis, rooted in our choice and use of different root metaphors, which separates humans from the more-than-human world. We propose that students spend time naming mechanistic metaphors that frame their daily lives as it enables them to move beyond simply raising environmental issues to a place where language plays a significant role in the development of anti-ecological habits of mind.

By way of another example and the work of Bowers (2003), we have also considered and examined Fritjof Capra's (1997) comparison of the mechanistic root metaphors embedded in the descriptions of a cell membrane, such as recycling centre, solar station, powerhouse, production centres, storage sacs, mini nucleus and nucleus, etc. They suggest that a plant cell is like a machine. Thus, we discussed how language emphasizes particular root metaphors that are embedded in thought patterns. Many TCs freely shared that they had never thought about a relationship between language patterns and the environment; the presentation of this concept at the workshop seemed to leave a lasting impression.

4.4 Teacher Candidate Perceptions of the Eco-mentorship Program and Learning Garden Field Placement

As part of the eco-mentorship program and learning garden field placement, TCs were asked to participate in a number of different curricular activities—workshops, curricula planning and development and community outreach—for the purpose of integrating local knowledges about the environment with their own emerging teaching identities. In order to help us discern how the importance of the planning and implementation of workshops and its resulting value was gained during the practicum placements, we used a mixed methodological approach to our data collection. Specifically, we collected data from a number of different sources, including reflective journals and diaries, interviews, written feedback and questionnaires.

In the end, the collected data helped us to see how the eco-mentorship program and learning garden field placement promote and develop a deeper understanding of how an analysis of root metaphors is integral to advancing a framework of best practice for integrated environmental learning. The following results about TC experiences in the eco-mentorship program and learning garden placement suggest how effectively prepared they are to be developers of environmental curricula in the classroom and beyond. Our research interests lie in the participant perceptions about their experiences as well as a particular focus on ecological literacy relating back to our eco-justice language curriculum. Following are some of the questions that the TCs were asked to consider: "What personal insights and/or reflections can you share about your experience in the eco-mentorship program and/or learning garden placement?", "What were some of the challenges you faced in relation to implementing environmental learning into your practicum and/or curriculum development?" and "How would you describe what it means to be eco-literate?" Responses from these questions were grouped into the following themes: ecological literacy and curriculum development and challenges to implementing environmental learning.

Under the theme of ecological literacy and curriculum development, the TCs expressed the following:

> Being aware of your footprint and impact on the environment. Leaving the world in a better state than when you enter it.
>
> Being able to go outside and recognise how it links to your learning.
>
> Awareness of the environment, how things grow, implemented into the classroom.
>
> Being able to critically think about the environment and how we are impacting it.
>
> Understanding how the environment is important and fragile – knowing proper practice in taking care of it.
>
> Learning garden is a hands on learning strategy that can be incorporated into all subject areas. Students should be taught about environmental education and food sustainability at all chances.
>
> It's important that students know where their food comes from and the role it should play in their everyday lives. They also need to learn that they can contribute and control how they eat, what they eat and how gardening can enhance their lives – both at school and at home.
>
> The school I was working at take pride in the appearance of their garden now and have said they will continue to work on it when I am gone. The school is now composting properly. Before, they put everything in one bin and there was plastic garbage found in it as well.

> In the front entrance of the school there is a giant collage of pictures that a teacher made of kids working in the learning garden next to their "eco-school" plaque. It also gives students a chance to be leaders in their school as well. Whenever I taught a class with students who volunteer in the learning garden in it, they had lots to say and seemed proud of their work. They wanted to take ownership of the garden and sometimes had to be reminded that the garden belongs to the whole school.
>
> I experienced nothing but positivity, encouragement, and gratitude for coming to help with a school garden. It seems to be something schools want to start but may not have the resources or the time to implement.

From these responses, we determined the following: the TCs see that awareness of the environment is important. Even more, we acknowledge that they believe human beings are connected to and influence the world at large, of which humans are always and already a part. Furthermore, the "outside," that is, "being outside," is something that affects one's learning about the environment. In other words, place matters and we have a role to play as we further understand how the environment is fragile and needs to be cared for through curriculum development.

We also found that some of our TCs expressed certain challenges in being able to engage in developing curriculum for environmental learning. For some, depending upon the context, making use of an eco-justice framework may continue to be challenging.

> The weather! I took my class outside and we created igloos, but the changeable weather interrupted the planned sequence.
>
> My Associate Teacher was not always on-board with going outside. She did not want to take responsibility for children outside. The focus was on math and language.

Of course, not all TCs experienced these types of challenges or issues. In fact, many shared experiences from their practicum placements that showed an openness to including learning gardens on school grounds and encouraging TCs to explore topics including sustainability and stewardship.

4.5 Finding the Educational Importance in an Eco-justice Framework

We believe that the educational importance of our study enforces the view that an eco-justice approach must also address language and should be considered as an effective method for TCs to enact ecological literacy in their classrooms. Furthermore, we also have seen how such an approach can foster the development of environmental curricula. From our experiences, enacting curricula that is environmentally focused is essential in building ecological habits of mind that can be passed on to younger generations.

> The development of ecological habits of mind involves the enhancement of an ecological consciousness that brings notions of interdependence, co-implicated relationships and the

> natural world into focus in daily educational practices that involve, among other things, literary explorations of root metaphors. (Young 2008, p. 1)

By reviewing the ways in which root metaphors, both mechanistic and ecological, play a role in our linguistic understanding about our shared connectedness to natural environments, TCs can help to foster ecological literacy through environmental an eco-justice education framework in K-12 education settings. It is our hope that our program's framework may be replicated across the globe and can promote sustainable environmental education practices that foster a more critical view of ecological literacy and greater leadership development.

Questions
Discussion questions that could help instructors and students to engage in meaningful conversation about the ideas presented in this chapter.

1. How might we consider the concept of root metaphors in relation to environmental education?
2. What skills and knowledge are required for environmental stewardship?
3. How might educators help foster the development of ecological habits of mind in students?
4. How do metaphors perpetuate systems of hierarchy?
5. How is language a barrier to our relationship with the natural environment?

References

Bell, N., Elliott, P., Rodenburg, J., & Young, K. (2013). Eco-mentorship: A pre-service outdoor experiential teacher education initiative at Trent University. *Pathways: The Journal of Outdoor Education, 25*(3), 14–17.

Bowers, C. A. (2003). *Toward an eco-justice pedagogy*. Retrieved from http://www.cabowers.net/pdf/anecojusticepedagogy2003.pdf

Bowers, C. A. (2004). *The eco-justice dictionary*. Retrieved from http://www.cabowers.net/CAdictmain.php

Capra, F. (1997). *The web of life: A new scientific understanding of living systems*. New York: Anchor Books.

Eco-Justice. (2008). *The Oxford English Dictionary*. Retrieved from http://www.britannica.com/EBchecked/topic/436518/The-Oxford-English-Dictionary

Longboat, D., Kulnieks, A., & Young, K. (2013). Beyond dualism: Toward a transdisciplinary indigenous environmental studies model of environmental education curricula. In A. Kulnieks, D. Longboat, & K. Young (Eds.), *Contemporary studies in environmental and indigenous pedagogies: A curricula of stories and place* (pp. 9–18). Rotterdam: Sense Publishers.

Martusewicz, R., Edmundson, J., & Lupinacci, J. (2011). *Eco-justice education: Toward diverse, democractic, and sustainable communities*. New York: Routledge.

Ontario Ministry of Education. (2009). *Acting today, shaping tomorrow: A policy framework for environmental education in Ontario schools*. Toronto: Queen's Printer for Ontario.

Orr, D. (1992). *Ecological literacy: Education and the transition to a postmodern world*. Albany: State University of New York Press.

Young, K. (2008). Ecological habits of mind and the literary imagination. *Educational Insights, 12*(1), 1–9.

Chapter 5
Embedding Environmental Sustainability in a Predominantly Online Teacher Education Programme: Ways to Contextualise Learning

Deborah Prescott

Abstract Through a teacher self-study, this chapter illustrates aspects of learning and assessment design in the context of a regional Australian tertiary institution with predominantly online delivery of 2 large (300–800 students) courses for teacher candidates. Four key guidelines for effective learning and assessment design have emerged through this investigation: focus on contextual development, collaborate for genuine purposes, embrace complexity, and build in reflexivity. These guidelines are fundamental to effective environmental education programmes as well as effective learning and teaching across all subject areas. They are illustrated in an environmental education scenario and are aligned within systemic sustainability education, a broad frame that encompasses complex systems, including the biosphere, which is not widely taken into account in teacher education. If these four guidelines are applied across the curriculum, environmentally responsive behaviours are more likely to develop because these guidelines are inextricably interconnected to transforming human society on a living planet.

5.1 Contextualising a Teacher Self-Study

A teacher self-study was conducted into learning and assessment design in the context of initial teacher education in a regional Australian tertiary institution with predominantly online delivery of large courses (300–800 students). The author was part of a team developing assessment and learning design in two curriculum theory courses in the Graduate Diploma in Teaching and Learning (GDTL). The approach we designed highlights the importance of developing context and accounting for complexity in learning and teaching. In addition, collaboration and reflexivity are

D. Prescott (✉)
Charles Darwin University, Ellengowan Drive, Casuarina Northern Territory, 0909 Casuarina, NT, Australia
e-mail: deborah.prescott@cdu.edu.au

G. Reis, J. Scott (eds.), *International Perspectives on the Theory and Practice of Environmental Education: A Reader*, Environmental Discourses in Science Education 3, https://doi.org/10.1007/978-3-319-67732-3_5

key pedagogical features also built into assessment and learning design. Pedagogical approaches used in designing these courses reflect characteristics of environmental education (EE) approaches.

Assessment and learning design in these two courses emphasise the process by which the content is conveyed to learners in a similar way as body language must be congruent with spoken words so they are more credible. "A most important component of the pedagogy associated with Education for Sustainability/Education for Sustainable Development is the process by which learners experience the material and values" (Thomas 2014, p. 1705) of EE.

Here, I show the pedagogical interconnections between our approach to two teacher education courses (on integrating literacy and numeracy in curriculum) and general approaches to EE. I argue that our team's approach to learning in teacher education is congruent with characteristics of EE programmes. Underpinning these similarities in pedagogies are frameworks from the Education for Sustainability area (e.g. Sterling 2012), from the teacher education area (e.g. LaBoskey 2004) and from both areas combined (e.g. Morin 1999).

The innovative pedagogical approach described here took place within a technicist paradigm, an educational worldview that values measurement and certification as well as traditional didactic pedagogies that promote competition, compartmentalisation, and control (Bamford 1999). Recent reviews by the National Department of Education and Training (2014) say there is too much complexity and too much prescribed content in the Australian curriculum and recommend that cross-curriculum priorities – such as sustainability – be taught only in relevant discipline areas, like Geography. This is counter to Edgar Morin's manifesto (1999) asserting that complexity and transdisciplinarity are essential for dealing with uncertainty and change in educating for a sustainable future.

In the Northern Territory of Australia, there are several factors that work against the traditional structure of teacher education courses (e.g. specialising in levels of schooling, discipline focussed, etc.) and traditional pedagogies. Firstly, school students' overall levels of achievement are significantly lower than the national profile, especially those in remote Indigenous communities. For example, in reading, the mean scale score for students in Year 7 (2015) was 483.4: the other states and territory had mean scores between 540 and 551 (ACARA 2015, p. 189). Secondly, 30 % of new teachers teach outside of their areas of expertise because of difficulties recruiting teaching staff (Australian Education Union 2009). Thirdly, greater use of online learning has contributed to teacher education at the author's university being multi-jurisdictional, including international students (McDowell 2013). These factors emphasise the importance of catering to diverse student cohorts for teacher candidates.

The predominantly online delivery of our courses meant that each student must make meaning in very different teaching and learning contexts around Australia (and sometimes internationally). More than 80 % of our students were external (Charles Darwin University enrolment data 2009–2015) and took the courses in mixed cohorts no matter where they lived (within the Darwin community or internationally), the school level or subject they preferred to teach, or the stage of their degree. A problem-based assignment task enabled divergent responses (where each student could contextualise their assignment response to their surrounding community) and widespread collaboration. Importantly, Peter McDowell (2013, p. 236) said about these diverse student cohorts that they "exit the unit [course] with a reasonable understanding of relevant curriculum and applicable literacy [and numeracy] teaching principles as these relate to their own particularised teaching contexts".

Illustrated in this chapter are some aspects of learning and assessment design that are likely to offer possibilities for transformative community action for students and promote deep learning, if enacted across their university degree, to challenge traditional assumptions about learning and teaching. Excerpts from assignment tasks demonstrate ways to help teacher candidates develop context, collaborate purposefully, and embrace complexity and reflexivity (Prescott 2016). I use an environmental education scenario to illustrate four key guidelines developed in accordance with Brent Davis et al.'s (2015) concept of systemic sustainability education (SSE). The notion of SSE started to emerge in the 1990s and is informed by the complexity sciences, ecology, systems theories, and decentralised network theory. The authors question the persistent paradigm of placing humans in a dominating role on the planet and reinforcing that role in formal education. I argue that the pedagogies we used in our courses were closely aligned to systemic sustainability education and were likely to develop ecologically literate and environmentally responsive citizens.

5.2 Environmentally Responsive Pedagogies

The environmentally responsive approach suggested here does not necessarily mean content from a discipline area, like science, nor does it imply content applied narrowly to a list of teaching methods (e.g. field trips). Assessment and learning design examples include phrasing of assignment tasks and activities that indicate broad pedagogical and process-oriented approaches drawn from context. Noel Gough (cited in Bamford 1999) asserted, "curriculum problems can only be resolved in light of complex human–environment interrelationships which must be treated holistically rather than analytically" (p. 170).

Over the years, while working on two literacy and numeracy curriculum courses with a team, it has become apparent to me that EE programmes use similar pedagogical approaches to those that we used: contextualisation (the importance of place and the grass roots), collaboration (e.g. purposefully working together to preserve habitats), embracing complexity (working within the transdisciplinary nature of the environmental issues and the tensions inherent in context), and reflexive practice (iterative, contemplative, recursively elaborative).

Alongside these pedagogical approaches, Davis, Sumara, and Luce-Kapler offer guidelines organised around an emerging and informed reconception of what teaching is. In their thought-provoking book, *Engaging Minds: Cultures of Education and Practices of Teaching* (2015), the authors outline helpful ways of viewing teaching within the frame of systemic sustainability education. They propose this term in order to reconceptualise teaching as "dancing of the designed and emergent"

(p. 218) to open interpretive spaces to solve complex, unanticipated problems, an "enlarging consciousness" to encompass the biosphere, and "recognising growing environmental concerns" (p. 169).

Many other authors raise questions about how the processes of learning that dominate current educational practices do not align to the processes of how students learn best or to the processes used in EE. For example, Greg Hunt (cited in Bamford 1999) notes that "changes in attitudes, values and behaviour inherent in environmental programs are difficult to reconcile with traditional assessment methods" (p. 164). Stephen Sterling (2012) summarises sustainability education as preparing people "to cope with, manage, and shape social, economic and ecological conditions characterised by change, uncertainty, risk and complexity" (p. 9). While the content that is included in our literacy and numeracy courses is important, so too is the process by which that content is conveyed to learners in order to help them adopt more responsive behaviours necessary to bring about environmental reparation. The processes of developing unique contexts and collaborating on a problem-based task were the same no matter if students were studying online or attending weekly classes on campus.

5.3 Dancing of the Designed and Emergent

One of the compelling metaphors in systemic sustainability education is that teaching is "the dance of the designed and emergent" (Davis et al. 2015, p. 218). The term "design" (rather than "plan") was used in our assignments to emphasise catering to diverse student needs, to enable contextualisation, to accommodate divergent purposes, and to make space for unanticipated, yet positive, possibilities to emerge. One of the assignment tasks in our courses was "to design and describe an effective learning experience that targets a small set of multimodal literacy needs as these needs arise in a context chosen by you" (2015 assignment). A task such as this offered "enabling constraints" (Davis et al. 2015, p. 219) – rule-bound parameters that allowed complex and unanticipated solutions to emerge as learning evolved. This task (and variations on it) allowed teacher candidates "variable entry" (Davis et al. 2015, p. 219) to the task so that they experienced success whatever context they drew upon. Learning design in our courses engaged all students in a manner appropriate to their current situation.

By combining notions from systemic sustainability education with the four guiding concepts for learning in teacher education (i.e. context, collaboration, complexity, reflexivity), we continued to reconceive effective learning design, underpinning not only EE but also all education. We continue to seek pedagogical approaches helpful in promoting learning overall – as well as environmental sustainability – in a diverse and primarily online student cohort in teacher education. The scenario that follows is illustrative of the EE programmes around Australia. Students can participate in similar place-based EE programs wherever they are living and studying and observe how the four pedagogical guidelines explored here are integrated into them.

Further, teacher candidates can offer these place-based EE learning experiences to their future students.

5.4 Environmental Education Scenario

Recently, I attended a community awareness excursion to a local area called the Howard Sand Plains. Of particular interest were endemic carnivorous bladderworts (*Utricularia* spp.) – some rare and endangered – that are under threat from the extractive industry, increased rural development, and recreational traffic (e.g. quad bikes). Greening Australia, a not-for-profit organisation dedicated to conserving and restoring landscapes, obtained funding to carry out studies on the species living in a shrinking and fragmented (but internationally significant and biologically diverse) area. Fifty participants attended the meeting, and there was a high level of engagement regarding action to be taken to protect these plants and their habitat (Fig. 5.1).

Being within the local context (this unique riparian environment) was essential to understand who the stakeholders were (even the adjacent golf course management was willing to collaborate with the Department of Natural Resources, Environment, the Arts and Sport, NRETAS) and what specific problems existed. The global context also became a factor because the Environmental Quality Report on Biodiversity of the Howard Sand Plains Site of Conservation Significance (NT EPA 2015) states, "bladderworts of the genus *Utricularia*... is of international significance" (p. 6) (Fig. 5.2).

Genuine collaboration was apparent in the community meetings convened by Greening Australia, where participants spanning a wide range of disciplines and

Fig. 5.1 The Howard Sand Plains (© Greening Australia)

Fig. 5.2 Example of *Utricularia* spp. (© Claire Elliott)

industries (including representatives from the sand mining companies) tried to work out precise problems as well as emergent solutions that would directly affect their community. The collaboration continued in the subsequent weeks and months: artists were commissioned to depict the Howard Sand Plains species, an education kit was published for schools (Greening Australia 2015), and an information booth was organised in the local community market.

The complexity of the issues was not lost on the participants. The open-ended and fluid processes evolved as new information and tensions came to light. For example, local experts on frogs and cane toads pointed out gaps in the Environmental Quality Report (NT EPA 2015). One of the *Utricularia* spp. experts found a particularly rare specimen, but, as careful as participants were, there was the real danger of trampling it. Another expert on the Howard River Toadlet – really a frog – described how little is known about its habits (it was only discovered in 2000) and how its common name, toadlet, contributes to sentiments against its conservation. (Cane toads, *Rhinella marina*, are destructive pests in Australia.)

As organisers and participants were moved to reflect on the interplay of factors impinging on *Utricularia* spp., key points were repeatedly encountered, reiterated, and revisited throughout the induction, the walk, and discussions. People articulated and clarified their understandings, reasoned and wondered aloud, retreated into solitary spaces, and formed networks.

These interactions occur in shared experiences around Australia and the world, wherever students are studying. Reflecting and collaborating through assignment tasks led students to confront the complexities of concerns in their context – some were of an environmental nature, others were centred on social justice issues, of which environmental concerns are a part (Morin 1999).

5.5 Pedagogic Parallels

In this section I build on and extend the EE scenario with the pedagogical guidelines of context, collaboration, complexity, and reflexivity to further illustrate our approach to two teacher education courses offered in both internal and external modes.

In my teaching experience, requesting a rich description of context in assignments required teacher candidates to include opportunities for listening keenly to their students, making informed responses to the school leadership, and tuning into the community priorities. When students described their teaching setting, there emerged complexities, tensions, and contradictions in their context. In Edgar Morin's book *Seven Complex Lessons in Education for the Future* (1999), he states, "we should develop the natural aptitude of the human mind to place all information within a context and an entity" (p. 2).

When teacher candidates undertake their professional experience in schools, lesson plan templates are widely used, so the way we phrased the assignment task – "design learning experiences" – caused students to pause and reflect on the difference between the two on the online discussion board. A lesson plan, when narrowly constructed in detail entirely in advance, is unlikely to allow reflexive teacher responses to the (possibly divergent and unanticipated) learning pathways that school students might take within a rich learning experience. Neither is a lesson plan sufficient for the teacher to adapt quickly to the "consciousness of the collective" (Davis et al. 2015, p. 221) of their students.

Environmental education – indeed, any topic (e.g. literacies and numeracies) – has little meaning until specific contextualities, and unique complexities are recognised as part of larger systems (Morin 1999). Inherent in this approach is that learning and teaching are not only "messy" but also provide many meaningful opportunities for learning. Effective learning does not lend itself to structured lesson plans, correct answers, competitive testing or linear thinking that dominates educational systems (Bamford 1999).

Jo-Anne Ferreira et al. (2014) are adamant that formal education institutions need to take a systemic view of change in order to sustain success in embedding EE in teacher education. They also emphasised that one of the success factors is the "congruence between the teaching and learning processes promoted and the principles of environmental education" (p.244), including dealing with uncertainty with regard to contextualised problems, and redefining complex relationships among various stakeholders – key aspects to purposeful learning. In the Howard Sand Plains scenario, "purpose" arose directly from the local citizens and community.

Promoting purposeful learning was a criterion for our assignment tasks. In a meaningful response to the task, "purpose" came directly from the context – students' preoccupations and curiosities, school, and community priorities. The quandary posed by this part of the assignment task – how do we know the learning is purposeful? – was a rich source of discussion of context (e.g. student questions, topics of school student talk, impending community events, and previous themes in learning).

The "dance" between salient parts of the context and new tensions and realisations arising from collegial discussion from other teaching contexts meant that students were practicing reflexivity (revisiting, reimagining, and reiterating) – a key message of the courses. The "dance" is also reflected in EE scenarios such as the one recounted about the Howard Sand Plains.

5.6 Enlarging Consciousness

My team and I designed our courses to create learning experiences that invite students to take on different, broader perspectives. Communicating with a contextualised audience was part of our learning design. For example, the structure and tone of a submission (in this case, an assignment task response) to the NRETAS minister about the Howard Sand Plains was completely different to a presentation for school students. When teacher candidates realised that they could choose a relevant audience in their context for their assignments, it was initially confounding because the audience for most university assignments, in our experience, was the lecturer. It also invited rich conversations on the electronic discussion board about how the structure of an assignment response reflected a particular audience in context. Students' choices of audience and purpose also had a significant bearing on their professional literacies and what genre (e.g. narrative) they chose to present, rather than defaulting to the standard academic essay. Choosing a meaningful audience helps students target their messages more explicitly and communicate more effectively in a professional context.

This is particularly relevant to this chapter and the literacy (and numeracy) learning emphases of our courses. Kimberly Moekle (2012) writes in the context of teaching composition to undergraduates "to explore matters of sustainability using the lens of rhetoric" (p. 77). She also explicitly connects the importance of genuine communication through literacy and the guidelines for environmentally responsive pedagogies as explored in this chapter:

> Rhetoric and composition [are] valuable precisely because the many challenges of sustainability will be solved not only through science and policy, but through writing studies that focus on communicative acts among various audiences... and to navigate the complexities of a planet on edge. (Moekle 2012, p. 83)

Prescribed audiences (an enabling constraint) for the assignments in these courses were pragmatic, but made room for various interpretations and contextualisation. For instance, students were asked to "take the position of a new educator sharing teaching ideas with colleagues" or "you are writing for yourself, online peers, and critical friends who can help you with your professional development" (2015 assignment). Students worked within broad problem parameters to define a targeted (but potentially infinite) audience. By way of illustration, the defined audience of "professional colleagues" could include other teachers, the Minister for Education, or the lecturer. The wider audience added dimensions and purpose to the assignment responses immediately applicable to students' current – or any other familiar – professional experience context.

We offered several suggestions to enlarge student consciousness of global environmental issues through assignment tasks invoking the Year for the Mathematics of Planet Earth (2013), the International Year of Pulses (2016), or the United Nations Decade of Education for Sustainable Development (2005–2014), for example. However, environmental content (e.g. Geography) is not a prerequisite for becoming more aware of earth stewardship.

Enlarging consciousness is often effectively achieved by creating disequilibrium within an assignment task (Davis et al. 2008, p. 106). We requested a "surprising idea or unanswered question", which provided particular insight into student contexts and motivated them to research further. Examples of surprising ideas were the extent of bullying or the lack of recycling at school. Morin (1999) exhorts us to "prepare our minds to expect the unexpected and confront it. Every person who takes on educational responsibilities must be ready to go to the forward posts of uncertainty in our times" (p. 13). When teacher candidates confront the unexpected and uncertainty in their contexts, it is highly likely to include social justice issues that are inclusive of environmental concerns (Morin 1999).

In these higher education courses, our learning design created opportunities for rich discussion and collaboration, opening up spaces for reflexivity and recursive elaboration. Some of the online activities made explicit suggestions to "divide up the task to save time", "discuss your findings with other students", and "articulate your collective understanding". In the course outline we stated, "online interaction forms an important means of students developing their professional language", "collegial learning is encouraged", and "student-to-student communication predominates". We underscored these statements by modelling collegial learning and professional language.

Online participation was an important contributing factor towards collaborative learning for external (and internal) teacher candidates. Widespread online interaction, however, was more successful when combined with an assignment task that requested students to "substantively improve another [teacher candidate]'s design for a learning experience". This task positioned students as knowledgeable colleagues and experts in their contexts. A remarkable collaborative feature of the cohort was that students spontaneously shared their assignments – whole or in part – before and after submission.

Students, positioned as experts in their learning and teaching context, started to challenge what tended to be taken for granted as normal (Davis et al. 2008, p. 180–1). Hierarchical structures (e.g. positioning the lecturer as the only expert in the course) are normalising whereas positioning students as experts called into question power differentials (Morin 1999, p. 24). Students started to constructively critique the curriculum, interrogate the status quo, and tentatively propose better approaches to learning in their context, including – but not limited to – environmental education.

Collaboration takes time and teacher educators are not always fortunate enough to have regular and frequent opportunities to genuinely collaborate on persistent patterns observed in learning in most tertiary environments. There are multiple and interacting factors contributing to undervaluing collaboration opportunities (for teacher–educators, teacher candidates, and school students). However, collaboration and collective action underpins systemic sustainability education. We made collaboration implicit (mentioned previously in this chapter) and explicit in assignment tasks by stating, for example, “it is vital to work collaboratively to share the workload and to further develop your critical perspective”. Students did collaborate with each other, but not because of prescribed group work – they did it voluntarily because it enlarged their consciousness.

Enabling collaboration among students is difficult if we ourselves do not collaborate in improving responsive learning design. Christine Edwards-Groves et al. (2014) observe:

> If our goal is to transform teacher practice then the answer lies in the provision of teacher professional learning in the context of learning communities that involve collaboration among teachers, students and outside experts…. There is a strong link between classroom talk, professional learning and the transformation of teacher practice. (p. 140)

Because of the high average student-lecturer ratio of 200:1 over these courses’ lives, we were not able to review draft assignments. This had unanticipated collaborative benefits: students decided to circulate their assignments on the Learning Management System if they wanted detailed, written feedback from their peers. The learning process thus invoked led students to evaluate each other’s assignments by using professional discourses, suggesting alternate ideas, identifying supporting resources, adapting learning design to their contexts, repeating key messages, and envisioning transformative education in context. It was very important for students and instructors to “validate teaching intentions through collegial discussion” (2014 assignment task).

Another way that we enabled enlarging consciousness was to build in reflexivity into our courses by linking the two assignments explicitly: Assignment 1 rehearsed Assignment 2 and Assignment 2 re-established the basis for Assignment 1. This approach offered a reflexive iteration, which, when applied in context, “respects and addresses the diversity, complexity, and contestation that’s present” (2014 assignment). The wording on the assignments indicated a possible direction students could take which would lead to a more desirable learning outcome for their students as well as themselves.

5.7 Making Environmentally Responsive Pedagogies Central in All Learning

Context is central in EE and must be accounted for in all learning and teaching. Collaboration is central to social action (e.g. environmental issues) and must be central to all education practices. Foregrounding complexity recognises the dynamic nature of learning and living systems. Reflexivity is built into learning design by emphasising the learning process over product and is also reflected in EE through critical enquiry.

The characteristics of these pedagogic approaches in our courses are congruent with environmentally responsive pedagogies that situate humans in the world – their common context – and not apart from it. Overall, "Education for Sustainable Development is about the kinds of education, teaching and learning that appear to be required if we are concerned about ensuring social, economic and ecological well–being, now and into the future" (Sterling 2012, p. 8). Morin (1999) put this in another way:

> The difficulty of knowing our World is aggravated by our mode of thought which has atrophied instead of developing the aptitude to contextualize and globalize, whereas in this planetary era we must conceive its globality, the whole–part relation, the multidimensionality, the complexity. (p. 31)

The Policy Studies Institute (2008) strongly acknowledges that "good sustainable development pedagogy is often simply good pedagogy" (p. 34). It goes on to recommend that higher education institutions engage more widely in educative processes which serve, over time, "to enhance society's understanding of what sustainable development might require through innovative and varied course content and pedagogy" (p. 35).

All educators contribute to individual and collective environmentally sustainable behaviours by applying environmentally responsive pedagogical approaches akin to the approaches we have applied, even in subject areas not traditionally associated with EE.

Questions

Discussion questions that could help instructors and students to engage in meaningful conversation about the ideas presented in this chapter.

1. Describe your context. Elements that you might include are:
   a. Your community's priorities
   b. Your passions and preoccupations and those of the people around you
   c. Your purpose for learning in this place at this time
   d. A topic arising in conversation: environmental sustainability, peace, equity, democracy, tolerance or a social justice issue
   e. What tensions and complexities exist
   f. Insightful direct quotes
2. "Students… started to challenge what tended to be taken for granted as normal". What would you challenge in your context?
3. Provide surprising ideas or unanswered questions you have encountered related to the context you described above.
4. Recount an environmental education experience that shows the importance of context, collaboration, complexity and reflexivity.

References

Australian Curriculum, Assessment and Reporting Authority (ACARA). (2015). *NAPLAN achievement in reading, persuasive writing, language conventions and numeracy: National Report for 2015*. Sydney: ACARA.

Australian Education Union. (2009). *New educators survey 2008: Results and report*. Southbank: AEU.

Bamford, B. (1999). From environmental education to ecopolitics. Affirming changing agendas for teachers. *Educational Philosophy and Theory, 31*(2), 157–173. https://doi.org/10.1111/j.1469–5812.1999.tb00382.x.

Davis, B., Sumara, D., & Luce-Kapler, R. (2008). *Engaging minds: Changing teaching in complex times* (2nd ed.). New York: Routledge.

Davis, B., Sumara, D., & Luce-Kapler, R. (2015). *Engaging minds: Cultures of education and practices of teaching* (3rd ed.). New York: Routledge.

Department of Education and Training of the Australian Government. (2014). *Review of the Australian curriculum –final report*. Retrieved from https://docs.education.gov.au/node/36269.

Edwards-Groves, C., Anstey, M., & Bull, G. (2014). *Classroom talk: Understanding dialogue, pedagogy and practice*. Newtown: Primary English Teaching Association Australia.

Ferreira, J., Ryan, L., & Tilbury, D. (2014). Planning for success: Factors influencing change in teacher education. *Australian Journal of Environmental Education, 30*, 136–146. https://doi.org/10.1017/aee.2014.39.

Greening Australia. (2015). *Secret world: Carnivorous plants of the Howard sand sheets*. Education Kit. Retrieved from https://www.greeningaustralia.org.au/uploads/related–downloads/SECRET_WORLD_Education_Kit__final.pdf

LaBoskey, V. K. (2004). Chapter 21: The methodology of self–study and its theoretical underpinnings. In J. J. Loughran, M. L. Hamilton, V. K. LaBoskey, & T. L. Russell (Eds.), *International handbook of self–study of teaching and teacher education practices* (pp. 817–869). http://doi.org/10.1007/978–1–4020–6545–3_21.

McDowell, P. (2013). Curriculum through literacy: Applying “parallel pedagogy” in pre–service teacher education. In A. Pandian, C. C. L. Liew, D. A. L. Tan, J. Muniandy, B. C. Lee, & C. H. Toh (Eds.), *New literacies: Reconstructing language and education* (pp. 230–246). Newcastle upon Tyne: Cambridge Scholars.

Moekle, K. R. (2012). The rhetoric of sustainability: Ecocomposition and environmental pragmatism. In K. A. Bartels & K. A. Parker (Eds.), *Teaching sustainability; teaching sustainably* (pp. 76–85). Sterling: Stylus Publishing.

Morin, E. (1999). *Seven complex lessons in education for the future* (N. Poller, Trans.). Paris: UNESCO.

Northern Territory Environment Protection Authority (NT EPA). (2015). *Environmental quality report: Biodiversity of the Howard sand plains site of conservation significance*. Retrieved from http://www.ntepa.nt.gov.au/about–nt–epa/comments/closed/howard–sand–plains

Policy Studies Institute, PA Consulting Group and Centre for Research in Education and the Environment (2008). *HEFCE [Higher Education funding council for England] strategic review of sustainable development in higher education in England*. University of Bath. Retrieved from: http://www.psi.org.uk/site/publication_detail/1089

Prescott, D. (2016). Telling reflections: Teaching sustainably in a complex learning environment. *Australian Journal of Environmental Education, 32*(1), 80–90. https://doi.org/10.1017/aee.2016.2.

Sterling, S. (2012). *The future fit framework: An introductory guide to teaching and learning for sustainability in HE*. York: The Higher Education Academy. https://doi.org/10.1177/0973408213495614b.

Thomas, I. (2014). Pedagogy for education for sustainability in higher education. *Sustainability, 6*, 1705–1708. https://doi.org/10.3390/su6041705.

Part II
Environmental Education Outside Walls

Chapter 6
Educating for Student Agency: Perspectives from Young Eco-civic Leaders in Canada

Lisa Glithero

Abstract This chapter shares the findings of a study that investigated how youth, nationally recognized as eco-civic leaders in Canada, perceive their agency and capacity to effect change. It explored the notion of "student agency" as it relates to the emergent trend around environmental action learning within the field of environmental education (EE) and civics education. Two key findings are examined: (1) the suggestion that a critical gap exists between student agency as interpreted in present school practices and scholarly and policy perspectives on EE and (2) the learning conditions that participating youth identified as critical in developing student agency, including the importance of youth–adult relationships. Our collective understanding and praxis of environmental action learning need to focus on the development (process) of students as active citizens, in the sense of collaborative civic actors aimed at socio-ecological change, not simply as "good stewards."

6.1 Moving Beyond Recycling Programs

Community cleanups, recycling and composting programs, school gardens and local food-based cafeterias, litterless lunches, no idling and water bottle-banning campaigns, and other behavior-changing initiatives are examples of student activism that have become commonplace in schools across Canada over the past decade (Astbury et al. 2009). This groundswell of environmental education (EE)-based activism is what practitioners and scholars alike frequently refer to as "environmental action" learning (Chawla and Flanders 2007, p. 440).

Specific to Canada, the current Ontario Ministry of Education's EE policy framework (2009) "Acting Today, Shaping Tomorrow" , for example, views students as "active citizens" (p. 13) and "decision makers to effect positive environmental change" (p. 15); teachers are asked to work toward building "student capacity to take action"

L. Glithero (✉)
Faculty of Education, University of Ottawa, 145 Jean-Jacques Lussier, Office 327, Lamoureux Building, Ottawa, ON K1N 9A7, Canada
e-mail: lisa.glithero@uottawa.ca; diz@CanadaC3.ca

G. Reis, J. Scott (eds.), *International Perspectives on the Theory and Practice of Environmental Education: A Reader*, Environmental Discourses in Science Education 3, https://doi.org/10.1007/978-3-319-67732-3_6

(p. 15) and "to effect long-term change" (p. 11); and "system leaders" (i.e., principals, administrators, policy-makers) are encouraged to make environmental leadership a "whole system responsibility" (p. 18). Similar trends toward environmental action learning have also been documented in recent curricular revisions including the Ontario Ministry of Education's Social Studies, History, and Geography (2013) curriculum document (grades 1–8), as well as several broader analyses of EE discourse focusing on sustainability and citizenship (Kozak and Elliott 2014).

Central to the above trend in EE discourse concerning environmental action learning is the recognition of the need to develop change agents in youth through the development of student agency, that is, building capacity in students to effect broader social change aimed at sustainability and socio-ecological well-being. The question remains, however: Can such "personal acts of responsible citizenship" (Westheimer 2008, p. 9) – the community cleanups, litterless lunches, water bottle-banning campaigns, etc. – drive the broader change being advocated for in current EE theory and policy? The study on which this chapter is based was designed to investigate how a subset of young people – specifically, young Canadians who have been recognized regionally and nationally as eco-civic leaders in their schools and communities – interpret their own sense of agency (i.e., their capacity to make change) and the kinds of learning conditions they perceived as formative in developing it. Although a limiting sample population, I was keen to examine the insights of youth who are engaged in environmental action learning processes taking place in schools or in local community structures as extensions of their formal learning.

This chapter is divided into two parts. In the first part, I examine the participating youth's perceptions of student agency in relation to the emergent trend of learning aimed at developing "citizenship skills" (Schusler and Krasny 2010, p. 210) and "action competencies" (Schnack 2000, p. 107) within the fields of EE. As for the second part, it presents a discussion on the learning conditions that these youth identified as critical in developing student agency. Before diving into these two central clusters of the ideas presented here, however, I want to briefly situate the relevance of this research within a broader international context.

6.2 Getting "Unstuck"

In investigating how youth, nationally recognized as eco-civic leaders in Canada, perceive their agency and capacity to effect change, a narrative inquiry approach was used in this research. An online questionnaire, semi-structured interviews, and a face-to-face focus group served as the primary data sources. Thirty-four past recipients or finalists of the Toyota Earth Day Canada Scholarship (TEDS) participated in it, and they were all aged 18 and 24 years.

Toyota Canada and the Toyota Canada Foundation, in partnership with Earth Day Canada, established TEDS in 2002 to help support emerging young Canadian environmental leaders. TEDS "encourages and rewards graduating Canadian high school students who have distinguished themselves through environmental community service, extracurricular and volunteer activities, and academic excellence" with a $5000 grant (www.earthday.ca). Each year the scholarship program recognizes 20 students, all of whom are entering their first year of studies at a Canadian university.

Although this chapter is based on a Canadian-centric study, the key findings discussed here bear upon discussions of EE pedagogy and praxis that are relevant to a much broader international audience. The research, along with my experience over the past 15 years as an environmental educator, has led me to the view that we are systemically "stuck" on an approach to environmental action learning that is focused on learning about (environmental) issues through knowledge building and awareness raising and opportunities to be good stewards and good citizens. Overlooked are deeper understandings and modes of learning that focus on processes aimed at collective action toward socio-ecological change (i.e., youth–adult codesign of a community sustainability strategy) and how youth can best co-participate in these processes. The perspectives of these youth participants – although Canadian – provide an important platform from which to provoke pedagogical conversation globally, enabling us as teachers, school administrators, scholars, and teacher educators to see environmental action learning differently and to become potentially "unstuck."

The currency of youth is their energy, innovative thinking, and inherent ability to see more relationally (Corcoran and Osano 2009). So, it is valuable to examine what solutions youth are already practicing or, at the very least, what information they might possess that can help us rethink and/or redesign EE praxis. Furthermore, the learning conditions put forth by the participants in this study offer insights from a youth perspective that might prove useful in engaging the traditionally disengaged youth – an important area requiring more research focus in the EE field.

6.3 Youth's Perceptions of Agency and the Coexistence of Ego/Eco-centric Thinking

An overwhelming majority of participating youth in the study (85%) interpreted the concept of "student agency" in relation to the idea of youth taking action toward positive personal and community change. In addition, 68% of youth are identified strongly as an eco-civic leader, although there existed a spectrum of interpretations as to what "eco-civic leadership" meant for them: one's "passion," one's level of "involvement in initiatives" or "engagement in issues" or efforts to "educate others," an overall "mindfulness of community needs," etc. Finally, 70% of participants perceived a strong personal capacity to effect change, although largely through individual behaviors and actions. In sum, a central tendency visible in the data is the inclination of these youth to perceive the locus of change and action to reside within the individual. In other words, most of these youth identified individual, personal traits such as motivation, interests, and passions and one's personal "actions" and "choices" as the vehicle for change. What does this first central finding – a widely distributed egocentric perception of agency – mean? Three points worthy of discussion are raised next.

First, these youth perceived school-based learning experiences as limiting their (development of) capacity for collective action, since the education system is not designed (although it is often imagined) as a space to challenge and change the status quo. As put by one participant: "our educational system doesn't empower students to take action; it's an institutional structure that offers little guidance on how to carry a sustainability mindset into a future career and civic life." In response, another two participants added: "there's just so many more skills we need to help us, like getting the ability to be the change makers" and "we need to witness and experience the process of making change." Schooling as a whole measures individual competency in a standardized body of knowledge and skills related to its application (Sterling 2009). In contrast, EE's focus on cultivating good stewards aligns with the individualistic nature of present public education, as pointed out by the study participants. Such egocentric thinking has informed how EE is pedagogically perceived and reinforced in the classroom. As expressed by yet another participant: "it's a shame how so much of my time outside of school, the institution that is supposed to prepare me for the world, doesn't… students should be educated to become key role players in society as they are growing up and should have such an integrated role in change occurring within society."

A second point of note is that the egocentric perception of student agency by these youths was, at times, accompanied by insights more reflective of eco-centric thinking. For instance, a relatively small number of youth (21%) perceived their capacity to effect change as "moderate" or even "limited." As one participant observed: "it is my intrinsic motivation, passion, interests, and determination that make me an eco–civic leader." She then went on to add: "the only way we're going to create change is through building our communities and this community–based approach." For youth like her, the importance of being a part of a community and feeling a relational sense of belonging "to something bigger than myself," including the more-than-human world (i.e., eco-centric or more ecologically oriented thinking), seemed necessary to cultivating, or perhaps legitimizing, their perceived sense of eco-civic identity.

Although seemingly contradictory – and perhaps reflective of the inherent complexities of our human condition – these particular youths' egocentric and eco-centric insights revealed a constructive coexistence; it was not a matter of either/or but rather that of and/both. For them, egocentric thinking is important as it serves as motivation, commitment, passion, and the idea that one's actions are pertinent. As suggested by one participant: "a young person needs to have it [motivation, commitment, passion] inside them, to say 'I'm going to do X'… this sense of wanting to achieve and succeed." Eco-centric thinking, on the other hand, matters as it represents a more reflexive understanding of participatory and collective action toward broader socio-ecological change (i.e., collaborative team member of developing a community-based water management strategy).

Closer analysis of the demonstrable coexistences of both egocentric and eco-centric thinking revealed another interesting correlation (Fig. 6.1). The more these

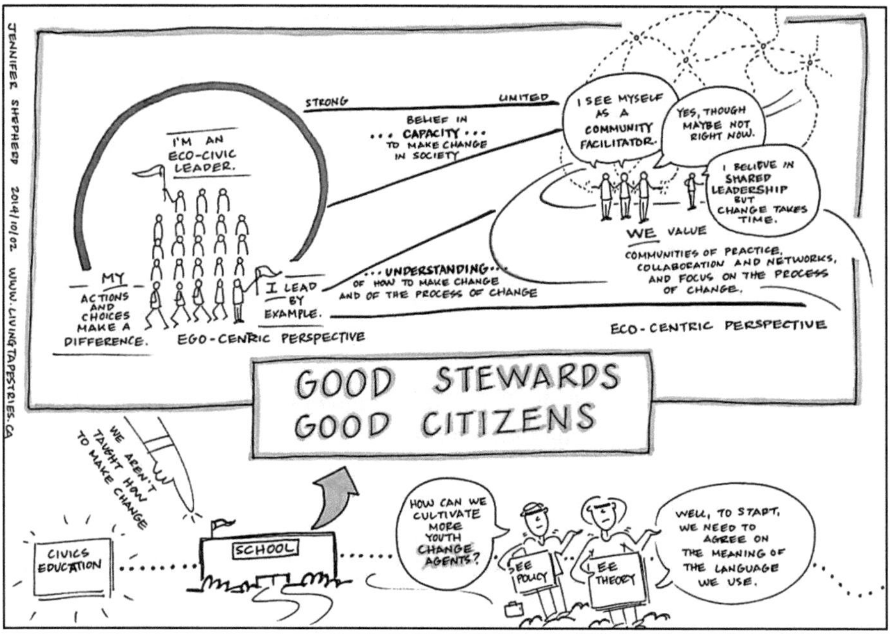

Fig. 6.1 Relationship between the individualistic lens (i.e., egocentric thinking) that was dominated among a large majority of participants in the study and the eco-centric lens that coexisted among a smaller subset. Visually represented is the striking correlation between a heightened understanding on how change is made (i.e., those few youth who spoke about "collaboration" and "communities of practice" [right side]) and their perceived "more limited" capacity to effect change (© Jennifer Shepherd)

youth perceived the locus of change and action to be in the collective sphere (i.e., what "we can do" as opposed to what "I can do"), the more limited their perception of the capacity to effect change. It seemed that a deeper understanding – or at least a heightened mindfulness – of the complexities of effecting broad social and systemic change has led these few self-identified as moderately to strongly eco-civic leader youths (9%) to hold a limited perceived capacity to make change. For example, one participant shared: "I am under no illusions about my own capacity; society is a massive, constantly evolving phenomenon with a huge amount of inertia… the approach that many people use is invasive and unsustainable." Perhaps for these youths, experience trumped idealism.

The third and final point I want to make with regard to the youths' perceptions of student agency is the argument that a critical gap seems to exist between how EE is widely practiced in schools with relation to active citizenship and how it is theorized in the EE literature. Generally speaking, mainstream EE practice in Canadian schools overemphasizes the individual as the locus of change and action through a narrow focus on fostering good stewards or engaging students in what I have dubbed "environmentally responsible stewarding" (Glithero 2015). However, as the emergent trend around environmental action learning (Arnold et al. 2009) and active citizenship (Astbury et al. 2009) critically suggests, along with more recent work linking EE to research focused on sustainability, citizenship, and change-oriented learning (Wals and Jickling 2009), the development of pro-environmental behaviors and personally responsive stewards, although important, is no longer enough. What is crucial is that we cultivate an active citizenry of youth who understand processes and are capable of participating in broader social and systemic change.

If EE practitioners are to embrace what current, progressive EE theory is advocating, then our collective pedagogical understanding of the locus of change and action in environmental action learning – or (environmental) education for "active citizenship" – needs to broaden and encompass not only the individual but also the collective and social. As the youth in this study demonstrated through their responses, it is not a matter of either learning for environmentally responsible stewarding *or* learning the knowledge and skills to effect broader social change; both are needed. Currently, we as EE teachers are generally fixated on the former (Capra 2007), while we need to embrace a broader understanding of change and a deeper understanding of environmental action learning experiences and of learning conditions that support students in developing the necessary knowledge and skills to collectively effect change. We need to move environmental learning beyond individual behavior change (i.e., personal acts of stewardship) to include (a) collective and "strategic behavior change" in the public realm (Chawla and Flanders Cushing 2007, p. 437) and (b) a deeper learned understanding of, and capacity for, making change on a social level.

For the youth in this study, schooling-based learning experiences, although formative in developing their motivation and behavior with regard to individual action around stewardship, still have room to develop their perceived awareness of and capacity to effect broader collective and social change. For the majority of these youth, it was those learning experiences that took place in the community, either through self-initiated volunteer, internship, and/or collaborative project building experiences, that helped support and/or develop their perceived sense of agency and change-making capacity.

Several youths in this study (36%) expressed needing, actively seeking, and/or being in the process of experimentally co-creating "supportive spaces of learning," something educational theorists might define as "communities of practice" (Wheatley 2005). Having identified such spaces as frustratingly void in public education, the youth found or built their own communities of practice. Examples include assuming a shared leadership role within a local food security and community resiliency building cooperative (http://nourishingontario.ca/true–north/); being a co-participant in a "living-learning community" focused on sustainability and social innovation (https://uwaterloo.ca/stpauls/residence/future–residents/environment–living–learning–community); co-founding a youth collaboration to address environmental, economic, and social challenges and opportunities in the Arctic (https://www.facebook.com/YouthArcticCoalition/); founding a community bike cooperative and building a network among other bike coops in the city (http://santropolroulant.org/en/what–is–the–roulant/collectives/santrovelo/); etc. In each of these instances, it was outside the school walls where these youth sought and developed the desired knowledge of, skills for, and opportunity "to do" collective action through collaboration.

As social movement theory and history itself have taught us, youth make up their own social movements (Clarke and Dougherty 2010). These often aim at challenging and changing the status quo or the system(s) associated with it. In Canada, in the past few years, for example, several youth-led social movements can be identified, including (a) We Canada (www.wecanada.org), (b) Powershift Canada (www.wearepowershift.ca), (c) Shannen's Dream (www.fncaringsociety.com/shannens–dream), and (d) Quebec students' march of 2012. In each of these cases, youth championed an innovative idea and mobilized thousands of other youths (and non-youths) around this idea through alternative approaches to traditional modes of social change (Clarke and Dougherty 2010). Much of the literature on youth civic engagement, youth social entrepreneurship, and youth leadership in social movements and social change occur, almost exclusively, in the public sphere (Zimmerman 2007). Youth-led movements typically occur in the wider public sphere, beyond formal schooling, because the conditions that allow for innovative ideas, and the risk-taking and mistake-making essential to creatively operationalize such ideas, are not only present but also widely encouraged in these environments (Johnson 2003). Moreover, youth-led social movements that both raise awareness and have impact in the public sphere are rooted in collaboration, partnerships, networks, and shared leadership models (Gauthier 2003), elements that are not as common in our century-old, industrialized schooling model.

6.4 Learning Conditions Important in Developing Student Agency

Turning to the second part of this chapter, I discuss the learning conditions that these youth identified as critical in developing student agency in the context of eco-civic action learning. The youths' insights on learning conditions focused predominantly on teacher praxis. A few insights also addressed broader schooling culture (e.g., shifting the perceived space of learning from classroom to community), as well as curricula more generally (e.g., developing certain skills like communication, presentation, and decision-making, among others). However, as discussed in the previous section, it was the learning experiences that they encountered when actively seeking to co-participate with peers and adults in community projects and processes aimed at broader social and environmental change (i.e., outside of the classroom) that best allowed them to put many of their insights into practice (i.e., communities of practice).

Specific to teacher praxis, the learning conditions that these youth collectively identified as formative shared several common features.

> These included the need for teachers to (1) make learning (i.e., curricula) more "personal," "relevant," and "interdisciplinary"; (2) embrace a more project, experiential, inquiry, and

> "action-oriented" approach to learning, one centered on students' interests; (3) perceive learning as "outputting," where experiences embedded in the local community enable students to apply their ideas, knowledge, and skills as active participants in real initiatives aimed at broader change; (4) value relationships, collaborative and participatory learning, and communities of practice; (5) serve as community connectors for youth and as models of active citizenship (i.e., to be more aware of opportunities and entry points for their students based on students' respective interests); (6) recognize identified skills (e.g., communication, partnership building, consensus decision-making, etc.) as important learning outcomes alongside a demonstrated knowledge of the curricula; and finally (7) recognize the importance of youth–adult relationships in which both serve as co-participants in the learning and change-making process.

In the next part of this chapter, I specifically focus on two of these learning conditions put forth above: (1) the idea of "learning as outputting" and (2) the importance of youth–adult relationships in which both serve as co-participants. These two conditions, I suggest, effectively serve to illuminate a necessary shift in schooling's approach to environmental action learning that may help us get "unstuck." By focusing on learning processes aimed at collective action toward socio-ecological change (i.e., youth–adult codesign of a community sustainability strategy), the following discussion contributes to a much larger and existing argument in the literature on the reorientation of teacher education to which I will address in the concluding section.

6.4.1 Learning as Outputting

The youth participants clearly expressed a desire for learning to be about "doing," to engage in the process of making change, and to have the "supportive framework," "partnerships," and/or "space" to implement "real initiatives that were actually happening out in the world…and not an exercise." Perhaps more pivotal for educators, these youth contextualized this as part of what formal learning (schooling) should be. The metaphor of school as "inputting" versus "outputting," put forth by Noah, a 20-year-old student, offers a useful perspective around which to shape further discussion of action-oriented learning:

> Students don't have to wait until they get out [of school] to start something… they are capable of making change when they are in school. My philosophy [on school] is that it is all about inputs… you getting knowledge, teachers are talking to you all the time, giving information to you but you are not outputting… you are not really contributing to anything to make a positive difference when you are in school. A lot of the times students feel that it is too much to do both at the same time and it does take time/skills but I do think young people are capable of outputting and making a difference while they are students. To create that vision a reality is it's an institutional change how education system is currently versus how we envision it to be… we need people at a local level who can drive programs that aim to and are dedicated to making all those connections between school and community and finding the avenues where they [students] can actually do stuff.

Noah, quoted above, self-identified the current approach of schooling as "inputting" and perceived this approach as a central impediment to developing his/students' capacity to engage in processes of effecting change in society – their agency. He went on to articulate a need for a shift in schooling culture (including teacher praxis), toward providing more opportunities for learning as "outputting," experiences that enable youth to participate in "real initiatives" in their communities and to authentically engage in the process of change. Noteworthy, Noah and other youth participants in the study appear aware on a very concrete level of what celebrated educational theorist Paulo Freire (1972) discussed much more abstractly with regard to the "banking model" of education – or the notion that the "scope of action allowed to the students extends only so far as receiving, filing, and storing the deposits" (p. 72).

In one illustrative example, Julie, age 18, identified the learning experience of building a community garden as "a powerful thing." In comparing it to school-based learning experiences, Julie went on to say: "if I'm talking about like concrete changes that I want, then doing them locally is going to, for me, be a lot more powerful than like letter writing campaigns [in school] where the results are very theoretical and out there." Beyond Noah and Julie, participating youth repeatedly identified community learning experiences that enabled them to embed themselves as participants in decision-making, policy-making, and project development initiatives that led to "meaningful" community-based change as their most formative learning experiences to date (e.g., "my civic–engagement work with Mississauga's Environmental Advisory Committee taught me about politics, how to get things done on committees, and a chance to actually make change"). For many of these youth, there was a stark contrast between the kind of learning that they experienced through various community participation and leadership opportunities and the "kind of learning in a class." Delving further into the 19-year-old youth's experience quoted above as a member of a local environmental advisory committee, Ben shared:

> Like it's day and night between learning in a class and then – I think for a lot of people, if they [as a class] went to a meeting at city hall and realized, you know that in the format within which things are discussed the range of issues that get brought up, they would see that, that they have a chance to actually make change, or that it's not as intimidating or difficult as they perceive, that would mean a lot more than just learning about different levels of government in class. Students need to witness the process of change.

We can glean from these participants' stories the imperative that more of the depth and quality of learning that they experienced across various community-based learning experiences need to be incorporated into the kinds of environmental action learning that could/should take place as part of public education learning. Inextricably linked to this point is the need to shift the locus of learning from the classroom to the community, a key condition identified by the youths themselves.

6.4.2 Youth–Adults as Co-participants

Some exciting work has been done on youth–adult relationships both within formal educational research and within broader public discourses pertaining to youth. Much of this work builds on Roger Hart's (1997) seminal "Ladder of Young People's Participation." Despite wide recognition and take-up of Hart's model, the enduring challenge highlighted by the experiences of participants in this study was how to cultivate such effective youth–adult partnerships in the context of formal education. The majority of youth in this study who addressed youth–adult partnerships developed these (trans) formative relationships outside of their school-based learning experiences. For one 18-year-old youth, her experience at a community-based organization in which she took the lead on writing a benchmark report on sustainability in her region served as a prime example. This report had been identified as a strategic priority for the organization, yet no one had "the time to take it on." Recognizing this, the youth in question, working as a volunteer, offered to take it up. Adult input was offered through subsequent team meetings, and the end result was a finished report that largely exemplified effective youth–adult partnership. In contemplating if such a partnership could take place within formal schooling, very few youth participants spoke about teachers giving them this kind of "freedom" and "support" or treating them as an "equally valued participant."

For the few youth who did speak about teachers as "peers" or "co-participants," what was common to the youths' descriptions of those teachers was the degree to which the teachers were embedded in their communities as actively engaged citizens. In addition, the youths identified these particular teachers as having a perceived awareness of myriad local, regional, and/or national opportunities for youths with which the students could connect. It seemed the challenge(s) for this kind of youth–adult (student–teacher) partnerships in formal learning experiences is not only a matter of the "structured" nature of schooling as an institution as aptly pointed about by the work of CEYE noted above. It is also a challenge, according to these youth, of teacher capacity, teacher education, and teacher identity (Iverson and James 2013).

The perspectives of these youth participants, particularly, with regard to the two learning conditions discussed above – the notion of learning as "outputting" and the need for youth–adults to serve as co-participants in collective social action-oriented project learning in the public sphere – provide an important platform from which to provoke pedagogical conversation on the need for teachers and teacher educators to see and "do" environmental action learning differently.

6.5 Teachers as Civic Actors

There are many conceptions of teaching – instructing, guiding, facilitating, mentoring, and empowering, to name a few. In his book, *Inventions of Teaching: A Genealogy*, Brent Davis (2004) offers a genealogical tree of contemporary conceptions of teaching. In doing so, he puts forth the argument that teaching, through an ecological epistemic lens, suggests teaching as "conversing, a quality of interpersonal engagement that is all but ignored in the traditional, radically individuated classroom" (p. 177). As Davis asserts, this concept of teaching "point(s) more to teachers' attitudes" than to teacher praxis, although I would argue that both are involved. Teaching through this lens directly resonates with, and builds upon, the related work of Nel Noddings' (1984) "ethic of caring," Max Van Manen's (1991) "pedagogical thoughtfulness," and Chet Bowers' (2002) "eco–justice," as Davis notes. He goes on to speak about teaching as "mindful participation" (p. 176) toward "expanding the space of the possible" (p. 179), drawing on Maxine Greene's (1995) work on wide-awakeness – "an awareness of what it is to be in the world" (p. 35) and the importance of "teaching for openings" (p. 109).

The insights shared by the youth participants in this study overlap with and extend (empirically) the work of Davis and related scholars in suggesting the need for teachers and teacher educators invested in environmental action learning to rethink professional learning around understandings and modes of learning that focus on processes aimed at collective action toward socio-ecological change (i.e, youth–adult codesign of a community sustainability strategy). This re-/orientation is a critically important condition for learning that is specifically aimed at youth engagement and student agency development. Part of the necessary reorienting work is about shifting teacher attitudes, as pointed out by Davis, to be "mindful in, being conscious of, being the consciousness of – the collective" (p.178). Another part of this work is about educating teachers on the value of becoming more aware and active participants in the communities in which they teach and live.

Although research has been done recently on "self-authoring a civic identity" (e.g., Iverson and James 2013), its focus has been on students and less so on teachers and teacher educators. More broadly still, this reorienting work is a cultural capacity-building effort, aimed at developing awareness of teaching and learning (and schooling) as an experiential and emergent process that takes place in community – as opposed to a standardized learning institution. This type of learning has to do with active, risk-taking participatory action (Reid et al. 2008). Small glimpses of this reorientation appeared within existing systems I examined, including the Ministry of Ontario's (Canada) "SpeakUp" and "Students as Researchers: Collaborative Inquiry Action Research" (i.e., www.edu.gov.on.ca/eng/students/speakup). However, to understand fully how far this reorientation has progressed and how to aid it, further more research is needed on learning experiences where students and teachers co-participate in action/change-oriented learning, as well as administrative support for such pedagogical experimentation (Reis and Guimaraes-losif 2012).

We need to orient EE practice, based on a wider understanding and adoption of promising learning processes emerging from such fields as social and participatory action learning and critical socio-ecological perspectives and from other works related to eco-civic action learning frameworks. Considering that "deep sustainability requires deep participation" (Wals and Jickling 2009), youths becoming young adults capable of participating as critical, democratic, political, social justice, ecological, and change-oriented citizens represent the learning endgame of educating for student agency.

Questions

Discussion questions that could help instructors and students to engage in meaningful conversation about the ideas presented in this chapter.

1. Name and discuss possible modes of learning that focus on processes aimed at collective action towards socioecological change.
2. To what extent do teachers need (or not) to be community activists in order to support the development of agency in their students?
3. How does teacher education need to shift in order to better support the development of student agency?

References

Arnold, H., Cohen, F., & Warner, A. (2009). Youth and environmental action: Perspectives of young environmental leaders on their formative influences. *The Journal of Environmental Education, 40*(3), 27–36.

Astbury, J., Huddart, S., & Theoret, P. (2009). Making the path as we walk it: Changing context and strategy on Green Street. *Canadian Journal of Environmental Education, 14*(1), 136–148.

Bowers, C. (2002). Toward an eco-justice pedagogy. *Environmental Education Research, 8*(1), 21–34.

Capra, F. (2007). Foreword. In A. Wals (Ed.), *Social learning towards a sustainable world: Principles, perspectives, and praxis* (pp. 13–15). Wageningen: Wageningen Academic Publishers.

Chawla, L., & Flanders Cushing, D. (2007). Education for strategic environmental behaviour. *Environmental Education Research, 13*(4), 437–452.

Clarke, A., & Dougherty, I. (2010). Youth–led social entrepreneurship: Enabling social change. *International Review of Entrepreneurship, 8*(2), 1–23.

Corcoran, P. B., & Osano, P. (Eds.). (2009). *Young people, education, and sustainable development: Exploring principles, perspectives, and praxis*. Wageningen: Wageningen Academic Publishers.

Davis, B. (2004). *Inventions of teaching: A genealogy*. Mahwah: Lawrence Erlbaum Associates.

Freire, P. (1972). *Pedagogy of the oppressed*. Harmondsworth: Penguin.

Gauthier, M. (2003). The inadequacy of concepts: The rise of youth interest in civic participation in Quebec. *Journal of Youth Studies, 6*(3), 265–276.

Glithero, L. (2015). *Exploring the development of student agency from the perspectives of young eco–civic leaders*. Retrieved from https://www.ruor.uottawa.ca/bitstream/10393/32335/1/Glithero_Elizabeth_2015_thesis.pdf

Greene, M. (1995). *Releasing the imagination: Essays on education, the arts, and social change*. San Francisco: Jossey-Bass.

Hart, R. (1997). *Children's participation, the theory and practice of involving young citizens in community development and environmental care*. London: Earthscan Publications.
Iverson, S. V., & James, J. H. (2013). Self–authoring a civic identity: A quantitative analysis of change–oriented service learning. *Journal of Student Affairs Research and Practice, 50*(1), 88–105.
Johnson, S. (2003). *Young social entrepreneurs in Canada*. Edmonton: Canadian Centre for Social Entrepreneurship.
Kozak, S., & Elliott, S. (2014). *Connecting the dots: Key learning strategies for environmental education, citizenship, and sustainability*. Toronto: Learning for a Sustainable Future.
Noddings, N. (1984). *Caring: A feminine approach to ethics and moral education*. Berkeley: University of California Press.
Ontario Ministry of Education. (2009). *Acting today, shaping tomorrow*. Toronto: Queen's Printer for Ontario.
Reid, A., Jensen, B., Nikel, J., & Simovska, N. (2008). Participation and learning: Developing perspectives on education and the environment, health and sustainability. In A. Reid, B. Jensen, J. Nikel, & N. Simovska (Eds.), *Participation and learning: Perspectives on education and the environment, health and sustainability* (pp. 1–18). Dordrecht: Springer.
Reis, G., & Guimaraes-losif, R. (2012). The death and life of a school–based environmental education and communication program in Brazil: Rethinking educational leadership and ecological learning. *Applied Environmental Education & Communication, 11*(3–4), 123–132.
Schnack, K. (2000). Action competence as a curriculum perspective. In B. B. Jensen, K. Schnack, & V. Simovska (Eds.), *Critical environmental and health education: Research issues and challenges* (pp. 107–126). Copenhagen: The Research Centre for Environmental and Health Education, The Danish University of Education.
Schusler, T., & Krasny, M. (2010). Environmental action as context for youth development. *The Journal of Environmental Education, 41*(4), 208–223.
Sterling, S. (2009). Ecological intelligence: Viewing the world relationally. In A. Stibbe (Ed.), *The handbook of sustainability literacy: Skills for a changing world* (pp. 77–83). Devon: Green Books.
Van Manen, M. (1991). *The tact of teaching: The meaning of pedagogical thoughtfulness*. London: Althouse.
Wals, A., & Jickling, B. (2009). A framework for young people's participation in sustainability. In P. B. Corcoran & P. Osano (Eds.), *Young people, education, and sustainable development: Exploring principles, perspectives, and praxis* (pp. 77–84). Wageningen: Wageningen Academic Publishers.
Westheimer, J. (2008). What kind of citizen? Democratic dialogues in education. *Canadian Education Association, 48*(3), 6–10.
Wheatley, M. (2005). *Finding our way: Leadership for an uncertain time*. San Francisco: Berrett–Koehler Publishers.
Zimmerman, K. (2007). Making space, making change: Models for youth–led social change organizations. *Children, Youth and Environments, 17*(2), 298–314.

Chapter 7
Environmental Education as/for Environmental Consciousness Raising: Insights from an Ontario Outdoor Education Centre

Joanne Nazir and Erminia Pedretti

Abstract Drawing on insights gained from a study of educators working together at an outdoor education centre in Ontario, Canada, this chapter aims to advance the idea that raising environmental consciousness involves connecting people to, fostering care for and building agency for their environments. Educating for the environment requires incorporating alternative education theories and providing people with personal, multidimensional experiences in the world, which are ultimately transformative. The chapter concludes with a discussion of the implications of these ideas for environmental educators and outdoor education facilities.

7.1 Setting the Scene

While environmental education (EE) has a longstanding history in Canada (Passmore 1972), a paper published at the turn of the millennium by Constance Russell et al. (2000) suggested that EE was by and large an inconspicuous or restrained endeavour. Since then, however, spurred by the general acceptance of human-caused environmental change, and in accord with many countries (e.g. Australia, Japan, United Kingdom and United States of America), EE has gained substantial interest across Canada.

Although EE has been an element of Ontario schooling (in different forms and degrees) over the last few decades, it was brought to the forefront of educational concerns in 2007 by the Ontario Working Group for Environmental Education's

J. Nazir (✉)
Redeemer University College, 777 Garner Road East, Ancaster, ON L9K 1J4, Canada
e-mail: jnazir@redeemer.ca

E. Pedretti
Ontario Institute for Studies in Education, University of Toronto,
252 Bloor Street West, Toronto M5S 1V6, ON, Canada
e-mail: erminia.pedretti@utoronto.ca

G. Reis, J. Scott (eds.), *International Perspectives on the Theory and Practice of Environmental Education: A Reader*, Environmental Discourses in Science Education 3, https://doi.org/10.1007/978-3-319-67732-3_7

report "Shaping Our Schools, Shaping Our Future: Environmental Education in Ontario Schools" (commonly referred to as the Bondar Report). This report unequivocally acknowledged environmental degradation "as a matter of increasingly urgent concern around the world" and described schools as having "a vital role to play in preparing our young people to take their place as informed, engaged and empowered citizens who will be pivotal in shaping the future of our communities, our province, our country, and our global environment" (p. 1). It urged immediate action to expand environmental education efforts and recommended the formation of partnerships among formal and informal sectors (e.g. schools, school boards, university research groups, environmental organizations and outdoor education centres), justifying these as necessary to providing environmental learning opportunities for all students. In the years since the release of this landmark report, several sectors within the province have responded to encourage EE in various ways. For example, Ontario's Ministry of Education created the document "Acting Today, Shaping Tomorrow" to serve as a policy framework for infusing EE into the province's schools (Ontario Ministry of Education 2009). It has also made efforts to translate these policies into practice by integrating EE into official curriculum guidelines for kindergarten to high school teachers (Ontario Ministry of Education 2011). Additionally, a number of non-formal educational organizations have strengthened their programmes and sought to establish stronger links with the formal education sector to provide diverse EE opportunities for students. For example, government ministries including Environment Canada and Natural Resources Canada have developed web-based educational resources that are freely accessible to educators and students, while private organisations such as Earth Rangers, Evergreen and EcoSchools provide EE programmes, mentorship programmes and learning resources for K–12 schools. A comprehensive list of these EE opportunities and efforts can be found in the document "Ready, Set, Green" (Ontario Ministry of Education 2007).

Yet, despite this flurry of activities, it is unclear how successful efforts have been in bringing about environmental teaching and learning. The few existing studies designed to explore the impact of EE in Ontario classrooms (e.g. Pedretti and Nazir 2014) suggest that a number of challenges – philosophical, cultural and practical – still hamper its successful pedagogical practice. Among other things, these studies note that teachers still find the following particularly troublesome to their EE efforts: understanding the relationship of EE to the rest of the curriculum, lack of pedagogical knowledge about enacting certain aspects of EE (especially action/activism) and limited opportunities for meaningful student interactions with the natural world. These findings indicate a problem that many scholars of EE have written about (e.g. Kahn 2008), that is, the need for forms of EE that take into account the complexities of classrooms, learners and environmental issues. In this chapter, we explore the consideration of EE as/for environmental consciousness raising and specifically through our work with EE at one outdoor education facility in Ontario, Canada.

7.2 Turning Ecological Ethics in Life Practice: Four Theories

Environmental consciousness raising is a term that has been used for some time now in EE forums (e.g. Krause 1993). While in the past it had been equated with knowledge and/or awareness of environmental issues (e.g. Santopietro 1995) or the performance of prescribed environmentally friendly behaviours such as recycling (Schlegelmilch et al. 1996), a review of more recent literature indicates that this is changing (e.g. Stevenson et al. 2013). Inherent to these newer conceptions is the understanding of humans as complex, existential beings of intentional consciousness and the concept of education as a process of change, which can only be brought about by multiple, personally tailored experiences. With this in mind, Arjen Wals and Justin Dillon (2013), for example, suggest that new constructions of EE should focus on transforming people from their existing ways of being to ways that support the long-term sustainable well-being of the Earth in all its fullness. Heesoon Bai and Serenna Romanycia (2013) capture this spirit of environmental consciousness raising when they say that EE is really about "turning ecological ethics into life practice" (p. 105) or, in other words, making ecological principles into habits of mind, body, spirit and heart. Supporters of this latter understanding of EE also point out that many mainstream programmes tend to focus on the cognitive, rational development of people, and to some extent emotional development, while other multidimensional aspects, that is, the physical and spiritual aspects of human consciousness, are largely ignored. Bronwyn Davies (2013) alludes to the significant embodied physicality of learning that traditional approaches to EE often ignore, while David Greenwood (2013) suggests that a significant aspect of raising environmental consciousness requires providing people with contextual embodied experiences, or simply put, deep mind-body-heart experiences. Bai and Romanycia (2013) further provide examples of how physical–spiritual exercises like mindfulness, yoga and meditation can work to bring people to environmental enlightenment.

Additionally, literature in the field of curriculum studies reveals a number of theories that are in sympathy with education for environmental consciousness raising as introduced above. In this section, we offer four of these theories for further consideration – holistic education, ethics of care, transformative learning and experiential learning with the belief that they will help the reader to gain a deeper understanding of what environmental consciousness raising means.

7.2.1 *Educating for the Whole Person*

Holism is the idea that the world is a seamless, dynamic, interconnected whole of which we humans are a part (Miller 2007). Holism accepts that humans are multidimensional beings with physical, mental, emotional and spiritual aspects, and that knowledge while complex is ultimately integrated into a seamless whole as well. The main goal of holistic education is to bring students, who often see themselves as fragmented, to a balanced, connected and inclusive relationship with the whole by educating the whole person. Many scholars have connected holistic education to EE. For example, John Miller (2007) describes nature as dynamic and connected and explains that an aspect of holistic education is to bring students to an awareness of nature's wholeness and their place in it.

7.2.2 Developing an Ethic of Compassionate Care

In 1977, Carol Gilligan disrupted the field of ethics by making the observation that emotion and intuition, rather than rational reasoning, is a foundation from which many people make ethical decisions. Gilligan (1977) defined ethics of care as deciding what is best for one's immediate relational group based on feelings of attachment and the need to maintain relationships. Nel Noddings (2002), one of the main scholars who have applied Gilligan's work to education, describes care-based education as relational, that is, making the themes of attachment, interdependence, the connected self and responsiveness to others central to the educative process. The focus is on educating the emotions – teaching students to be competent carers and sensitive cared – so that they can conceptualize problems in a more compassionate way and genuinely seek to find out what is best for all in a particular situation. It is important to note that ethics of care go beyond building rationally justifiable affection or concern and in so doing offer a different axiological base upon which people can build different relationships with environments, entities and non-human others. Applied to EE, several scholars (e.g. Martin 2007) suggest that students should be encouraged to recognize the Earth as a living entity who sustains human life and in turn respond to form a relational bond with her, as one would with a loved Other.

7.2.3 Experiential Learning

Theories of experiential learning emphasize the personal and active processes of coming to know. According to David Kolb (1984), experiential learning consists of four processes: concrete experience with phenomena, observation and reflection, forming of new knowledge and the application and testing of new concepts in new situations. While all four processes are important, for many advocates, the focus of experiential learning is on concrete, personal experience with phenomena (e.g. Breunig 2008). Over the years, distinguishing educative experience from superficial experience has been a major thrust. In this regard, Clifford Knapp (1992) suggests educative experiences should involve meaningful, challenging interactions with phenomena in real-world settings, while Peter Higgins and Robbie Nicol (2011) state that experience must involve multidimensional engagement of the mind, emotions and senses. Experiential learning has found expression in EE most clearly in the about-in-for model of EE (Lucas 1979). This model suggests that direct contact/

experience with surroundings is essential to students' environmental learning by bringing about appreciation for nature and providing concrete experiences of doing environmentally friendly actions.

7.2.4 *Transformative Learning*

Supporters of transformative learning insist that more than passing along facts, true learning should bring about deep changes that move people to new and positive ways of being. For example, Edmund O'Sullivan (2003) states, "Transformative learning involves experiencing a deep, structural shift in the basic premises of thought, feelings, and actions. It is a shift of consciousness that dramatically and irreversibly alters our way of being in the world" (p. 237). Applied to EE, this understanding of transformative learning suggests that pedagogy for environmental learning should focus on bringing about learner transformation rather than passing along facts and skills.

While theories of holistic education, ethics of care, experiential learning and transformative learning differ in their individual details, they share certain underlying premises. They all suggest that human nature is complex, existential and intentional and that environmental educators need to take these characteristics of the human condition into account in their conception and practice of EE. In our study of outdoor educators' work with EE, we found these theories particularly salient in interpreting educators' understandings and activities with respect to EE and in coming to the conclusion that they are working from an attitude of EE as environmental consciousness raising (as introduced earlier in this chapter).

7.3 The Uneasy Relationship Between Outdoor and Environmental Education

This study of EE occurred within the context of an outdoor education centre (OEC). For many, the kinship between outdoor education and environmental education is obvious. Russell et al. (2000) describe outdoor/nature-based programmes as one of "the diversity of narratives" (p. 207) through which EE is practiced in Canada. Lucie Sauvé (2005) another well-established Canadian environmental scholar states:

> Outdoor education is one of the most effective means of learning about/within the natural world and imparting an understanding of nature's inherent right to exist by and for itself–humankind's place in nature being definable only in context of this ethos. (p. 14)

Andrea Foster and Grant Linney (2007), speaking on behalf of the Council of Outdoor Educators of Ontario (COEO), assert:

> Early sequenced and repeated experiences in the outdoors [help students] develop a kinship with nature that can evolve into an informed, proactive, and lifelong stewardship of our natural environment. (p. 53)

Yet, despite such positive popular endorsement, a thorough review of the scholarly literature suggests that the relationship between outdoor education and EE is not as straightforward as some imply. Indeed, there are scholars who doubt the compatibility between the two fields. Chris Loynes (2002) and Annette Gough (2007), for example, argue that much of what occurs in mainstream outdoor education is outdoor skill development, physical fitness and interpersonal development, and that the activities associated with these promote environmentally detrimental rather than environmentally positive relationships with nature. Furthermore, according to Loynes (2002), in mainstream outdoor education "nature is understood as an assault course, gymnasium or puzzle to be resolved and controlled. It is a resource to be commodified instead of a home to which to relate" (p. 3). According to Alison Lugg and Dierdre Slattery (2003), in many mainstream outdoor education contexts, EE is limited to minimum impact behaviour and aesthetic appreciation of outdoor settings.

As much as these criticisms concerning the compatibility between outdoor and EE may be true for some contexts, they are at variance with our understanding and experience of environmental and outdoor education. Outdoor education programmes in Canada have long been associated with EE (Russell et al. 2000). Further, in a previous research study we conducted (Pedretti and Nazir 2014), Canadian educators expressed a strong belief that there is a significant overlap between outdoor and environmental education and expressed a longing for the provision of more outdoor opportunities to engage students in environmental learning. In a bid to further understand the relationship between outdoor and environmental education in the Canadian context, we conducted this study of outdoor educators' work with EE at Faraway Dale Outdoor Education Centre (pseudonym). We believe the analysis and results of this study will be of interest to environmental and outdoor educators anywhere, both theoretically and practically.

7.4 Doing Environmental Education at Faraway Dale Outdoor Education Centre

Faraway Dale Outdoor Education Centre is located in a large city in Ontario, Canada. It is owned and operated by a major provincial school board. The facility, which has been used for outdoor educational purposes for over 50 years, consists of 55 ha of natural green space encompassing meadow, forest, marsh and river habitats. Faraway Dale is a day centre (as opposed to a residential centre) and serves the urban and suburban student population within its vicinity. It accommodates over 1,700 elementary-aged students from 150 schools on visits throughout the school year. Faraway Dale has been mandated by its parent organization to provide EE as part of its educational efforts. However, other than minimal guidelines, it has been left to educators at the facility to interpret what this means and develop appropriate pedagogy to fulfil this mandate. In response, over 50 modified/self-generated programmes (e.g. Earth Repair Projects, Hiking in Nature, Measuring Biodiversity, Exploring Textures in Nature, Physical Patterns in the Changing World, Snowshoeing, Collaborating with Nature and Habitats and Communities) are offered at Faraway Dale. The overarching principle governing programmes is that they should provide cohesive, impactful sequences of outdoor activities to meet the centre's mandated purposes, including EE.

We chose to study the nature of EE at Faraway Dale Outdoor Education Centre by examining the lived experiences of nine educators who had each worked there at the time for at least 10 years. We refer to the four men and five women who participated in the study with the following pseudonyms: Bruce, Danny, Keith, Trevor, Carol, Kelly, Nesha, Arlene and Ellen. Data collection took place over a period of 5 months during which we visited the centre regularly on daylong visits at least twice every week. Four data collecting strategies were used on these visits: two semi-structured interviews with participants, field notes of outdoor education sessions, participant journals, and collection of relevant artefacts (e.g. documents, pictures, and memorabilia); at the end of data collection, we had over 500 pages of documented data. Next, we applied established qualitative data analysis procedures suitable for analysing human experience (e.g. Polkinghorne 1989) to illuminate the nature of EE at the centre. Environmental education as/for environmental consciousness raising emerged as an overarching theme of this analysis. Participants' experiences provide deeper understandings of what this concept is and how it can be translated into pedagogical practice.

7.4.1 Educators' Insights About Raising Environmental Consciousness

At the beginning of the research process, we asked educators at Faraway Dale the following question: "What is Environmental Education?" Their responses suggest that they were working from an understanding of EE as environmental consciousness raising, with an emphasis on turning ecological ethics into practice or action:

> [EE] is when your actions support what you say... that you walk your talk. It means I am bringing in my mug, not a cup that I am going to throw away every day. (Nesha, Interview 1)

> ...to me it's [EE is] about internal changes. It's not about don't pick and pull the branches because we are watching you. It's about... because you feel a connection for the tree. And then it doesn't matter where [students] are. They're not going to do it because they care and see the harm they might cause. (Arlene, Interview 1)

Talk about "walking your talk" and not harming nature because "someone is watching you" indicates that for these educators, EE is about transforming people intrinsically and nurturing more relational interactions among people and nature rather than simply bringing about prescribed, extrinsically motivated actions. Additionally, comments such as "feeling a connection" demonstrate educators' commitment to environmental consciousness raising as opposed to simply passing along the rational canons of knowledge about the living world.

As the research process continued, Faraway Dale educators' theoretical and practical understandings of EE as consciousness raising became clearer to us. These took the form of three salient structures: connecting people to the environment, fostering compassionate care for the environment, and building agency for living low consumption life styles. Below, we expand on each of these structures and illustrate how they are enacted through exemplar pedagogical practices.

7.4.2 Connecting People to the Environment

For these educators, connecting refers to the formation of a personal, tangible, multidimensional bond with nature. Furthermore this bond is positive, natural and comfortable much like the bond that ties a mother to an unborn child. Consider the following comments concerning the meaning of connecting to nature:

> Connection means finding your place. I think it means seeing where you belong. And that's not about looking down and going, "Here I am and there is everything". But standing where you are and going, "This is my world. This is the air I breathe. This is the water I drink." And knowing there is a whole world out there that you can live in, and be, and prosper–and it doesn't involve money and status. It involves touching, and smelling, and looking and feeling in your heart, "This is where I belong". (Keith, Interview 1)

> You can teach about the natural world and how it works, but connecting is more than that. It's that peace and being within yourself that nature provides in a unique way. (Trevor, Interview, 1)

An example of how connecting with nature is encouraged at Faraway Dale is through the Hug-a-Tree activity. This activity takes place during a hike in a forested area. As researchers, we were fortunate to witness it on several field visits with different educators. On these occasions, Hug-a-Tree usually occurred after students had learned how to identify a tree. Students were simply invited to identify a particular tree and give it a hug. While some students were initially sceptical about hugging a tree, on each occasion we saw it, those that ended up doing it seemed pleasantly gladdened having done so. As Trevor offered, "In some ways tree–hugging is a bit silly. But at the same time you're hugging a tree. It's intimate and you're connected to it. You open yourself up to something when you do that" (Trevor, Field Visit 1). For us, the activity highlights these educators' understanding of environmental education as more than a rational cognitive process – in this case the need to physically touch as part of the process of coming to know. Pondered more deeply, educators' comments and the pedagogical example suggest that connecting is really about nurturing a holistic relationship with the world, which requires multidimensional, authentic experience in the world as suggested by holistic education and experiential educational theories discussed earlier in the chapter.

7.4.3 *Encouraging Compassionate Care*

Based on Faraway Dale's outdoor educators' practices, experiences and responses, it seems that they believe that part of their job is to encourage the development of compassionate carers and sensitive cared for:

> I think it's important for [the students] to understand how to care for nature and the Earth as they would for a family member. (Carol, Interview 1)

> Faraway Dale is a place for students to encounter nature face–to–face, brain–to–brain and heart–to–heart so they can connect and build a caring relationship with her. (Arlene, Journal)

Further, their understanding of care is highly reflective of Noddings' (2002) ideas of care-based education outlined earlier. Care is encouraged amongst students and provided to students by educators. Applied to EE, educators encourage visiting students to recognize nature as a living Other that sustains human life and to respond sensitively and positively to form a caring relationship with her. Moreover, students are not expected to rationalize why they should care but rather to do so naturally from a place of emotional depth.

Educators at Faraway Dale infuse the notion of care across programmes in several ways. At the beginning of each educational session students are explicitly introduced to care as an appropriate stance to adopt toward human and non-human others while at the centre. Educators also encourage care by modelling it themselves. With students they are sensitive, gentle and responsive. Out in the environment, they are careful not to step on saplings and tree roots, stick to pathways so as to be as undisruptive as possible, remove bits of garbage they find in the forest and handle wild animals only if they need to. They also overtly recognize and praise instances when students demonstrate care. Across the research process, we came to realize that the notion of compassionate care may be one of the fundamental ideas about the overall nature of education that educators in this context hold. Keith made this clear in explaining his educational philosophy: "We all need to feel like we're something and that we are cared for" (Keith, Interview 1).

7.4.4 *Building Agency*

The third structure of environmental consciousness raising that emerged from our work with educators at Faraway Dale was building agency for the environment. For these educators, building agency was crucial and involved providing concrete experiences of interacting with the environment in ways that are more benign. According to one of the participants:

> Most environmental programs are naked of the fact of how to act for the Earth. Don't do this and don't do that rather than turning it around to show what we should be doing. (Carol, Interview 1)

Faraway Dale educators always include one agency building activity as part of the programme with every visiting group to the centre. One such activity was the Earth Repair Project (ERP), which allows students to participate in a concrete way to ameliorate human caused damage on the facility's grounds. An ERP can take many forms, for example, restoring pathways, planting butterfly or medicinal gardens, controlling invasive species, and afforestation. During data collection, we observed restoring pathways through woodchipping on several occasions. For this activity, students are brought to an eroded pathway. Here, they observe degradation of the pathway. The method of resurfacing the path with a layer of woodchips is demonstrated. Students are then provided with simple equipment (e.g. buckets, shovels and a pile of woodchips) and tasked with resurfacing a given length of pathway. Next, they are invited to experience the resurfaced pathway and share their thoughts and feelings (both positive and negative) about the activity. According to educators at Faraway Dale these concrete, interactive experiences scaffold students in their movement from wanting and thinking about acting for the environment, to feeling empowered to do so, if they are so inclined. In Danny's words:

> Having them [the students] do woodchipping on a path helps them to feel that they are part of something greater, something beyond themselves. And it teaches them to look for instances where they can help to take care of the Earth around them. (Danny, Fieldnotes from Field Visit 2)

These ideas resonate with those described in experiential and transformative learning theories. The hope is that visitors are experiencing more authentic forms of EE (Kahn 2008) and are slowly beginning to alter their ways of being in the world (O'Sullivan 2003).

7.5 Looking Forward

It is often lamented in Ontario (e.g. Pedretti and Nazir 2014) and elsewhere (e.g. Rickinson 2001) that despite much effort the efficacy of existing EE programmes is dubious. Over the years, partly in response to this problem, the nature and practice of EE have been questioned. Many have asked what it means to be environmentally educated and have come up with various responses. While recent formulations of EE tend to focus on getting people to think deeply and critically about environmental concepts and issues (e.g. Kahn 2008), others have suggested that these, along with other forms of EE, stem from an incomplete view of people as primarily cognitive-rational or mainly thinking entities (e.g. Davies 2013). Where other dimensions such as emotion and physical action are included in the educative process, they are often treated as subservient to thinking. Indeed, a popular underpinning framework for many forms of EE is the know-concern-act model which assumes that knowing about the environment in a technical/rational way will lead to an appropriate feeling of concern for it, which will in turn act as a motivator for positive environmental actions (e.g. Hungerford et al. 1983). In agreement with those who see these assumptions as problematic, we support a formulation of EE that takes into account the complex aspects of human nature, particularly the notion that humans are beings made up of mental, physical, emotional and spiritual aspects and that all of these must explicitly be taken into account in any EE endeavour. We suggest that EE as/for environmental consciousness raising as we have described at the beginning of this chapter embraces these ideas. Further by linking the ideas of environmental scholars to relevant curriculum theories (i.e. holistic education, ethics of care, experiential learning and transformative learning) we have endeavoured to strengthen the theoretical base for viewing EE as/for environmental consciousness raising.

Further, we suggest that the work of outdoor educators at Faraway Dale Outdoor Education Centre advances our understanding of what this type of environmental consciousness raising means in practice. It does so in two ways: (1) by unpacking the term into the structures of connecting, care and agency and (2) by providing concrete examples of pedagogical activities for accomplishing these structures. Connecting, care and agency inherently take into account the complex nature of humans and the interpretation of education as a process of transformation which seeks to bring about deep changes in consciousness (O'Sullivan 2003). More than cognitive-rational knowing, connecting implies the formation of a personal, tangible bond with nature. More than concern, care for environments arises out of an ethic of compassion and originates from a place of emotion and spirit. Rather than prescribing actions for the environment, agency is about empowering students by building the motivation and skills to act or not act in ways they personally choose. In an outdoor context, accomplishing these structures translates into pedagogical practices that provide personal, contextually relevant, multidimensional experiences (i.e. involving mind, body, emotions and spirit) in the real world.

Conceptualizing EE as a process of this type of environmental consciousness raising has implications for different audiences. For many environmental educators and researchers, it will require a shift in ideology with respect to the nature of the human being and the educative process. To facilitate this shift, what may be most needed is personal experience of learning that not only emphasizes the cognitive rational aspect but also honours learning through the body, emotions and spirit, similar to what we have described as common practice at Faraway Dale Outdoor Education Centre. In other words, educators may need to experience alternative educational processes and be transformed themselves before they can begin to fully understand the nature of environmental consciousness raising and pedagogies that support it.

For outdoor education workers and its supporters, this notion of EE as/for environmental consciousness raising also opens an exciting avenue. It establishes more concretely the link between outdoor and environmental education by demonstrating how education in the outdoors is necessary to environmental learning by providing powerful educative and transformative experiences. It provides outdoor educators with structures and theories they can work with as they continue to develop their own EE programmes. This is important especially in a time when the question of relevance is an issue that existing outdoor education centres are increasingly being asked to respond to. One strong response that such facilities can make is that they are essential for providing students with the kinds of personal, multidimensional experiences with the natural world required to build connections, care and agency for the environment and ultimately raising students' environmental consciousness.

Questions

Discussion questions that could help instructors and students to engage in meaningful conversation about the ideas presented in this chapter.

1. In "setting the scene", what underlying problem(s) do the authors acknowledge as their impetus for writing the chapter? Do these problem(s) resonate in your situation and if so, how?
2. What are the salient features of "environmental consciousness raising" according to the authors?
3. How does the reported study of Faraway Dale Outdoor Education Centre deepen your understanding of environmental education as/for consciousness raising?
4. How might an environmental educator in a non-outdoor setting include the notion of consciousness raising into her/his pedagogical practice?

References

Bai, H., & Romanycia, S. (2013). Learning from hermit crabs, mycelia and banyan: Schools as centers of critical inquiry and renormatization. In R. B. Stevenson, M. Brody, J. Dillon, & A. E. J. Wals (Eds.), *International handbook of research on environmental education* (pp. 101–107). New York: AERA Routledge.

Breunig, M. (2008). The historical roots of experiential education. In K. Warren, D. Mitten, & T. A. Loeffler (Eds.), *Theory and practice of experiential education* (pp. 77–92). Boulder: Association of Experiential Education.

Davies, B. (2013). A feminist post-structural approach to environmental education research. In R. B. Stevenson, M. Brody, J. Dillon, & A. E. J. Wals (Eds.), *International handbook of research on environmental education* (pp. 438–458). New York: AERA Routledge.

Foster, A., & Linney, G. (2007). *Reconnecting children through outdoor education. A research summary*. Toronto: The Council of Outdoor Educators Ontario.

Gilligan, C. (1977). In a different voice: Women's conceptions of self and morality. *Harvard Educational Review, 47*(4), 481–517.

Gough, A. (2007). Outdoor and environmental studies: More challenges to its place in the curriculum. *Australian Journal of Outdoor Education, 11*(2), 19–28.

Greenwood, D. (2013). A critical theory of place conscious education. In R. B. Stevenson, M. Brody, J. Dillon, & A. E. J. Wals (Eds.), *International handbook of research on environmental education* (pp. 93–100). New York: AERA Routledge.

Higgins, P., & Nicol, R. (2011). Sir Patric Geddes, "Vivendo discimus" – by living we learn. In T. Smith & C. Knapp (Eds.), *Sourcebook of experiential education: Key thinkers and their contributions* (pp. 32–40). New York: Routledge.

Hungerford, H. R., Peyton, R. B., & Wilke, R. (1983). Editorial– Yes EE does have a definition and structure. *Journal of Environmental Education, 14*(3), 1–2. https://doi.org/10.1080/00958964.1983.9942655.

Kahn, R. (2008). From education for sustainable development to ecopedagogy: Sustaining capitalism to sustaining life? *Green Theory and Praxis: The Journal of Ecopedagogy, 4*(1), 1–14. https://doi.org/10.3903/gtp.2008.1.2.

Knapp, C. (1992). *Lasting lessons: A teacher's guide to reflecting on experience*. Charleston: ERIC Clearinghouse on Rural Education and Small Schools.

Kolb, D. A. (1984). *Experiential learning: Experience as a source of learning and development*. Saddle River: Prentice Hall.

Krause, D. (1993). Environmental consciousness: An empirical study. *Environment and Behavior, 25*(1), 126–142. https://doi.org/10.1177/0013916593251007.

Loynes, C. (2002). The generative paradigm. *Journal of Adventure Education and Outdoor Leadership, 2*(2), 1–14. https://doi.org/10.1080/14729670285200221.

Lucas, A. M. (1979). *Environment and environmental education: Conceptual issues and curriculum implementation*. Melbourne: Australia International Press and Publications.

Lugg, A., & Slattery, D. (2003). Use of national parks for outdoor environmental education: An Australian case study. *Journal of Adventure Education & Outdoor Learning, 3*(1), 77–92. https://doi.org/10.1080/14739670385200261.

Martin, P. (2007). Caring for the environment: Challenges from notions of caring. *Australian Journal of Environmental Education, 23*(1), 57–64. https://doi.org/10.1017/S0814062600000719.

Miller, J. P. (2007). *The holistic curriculum* (2nd ed.). Toronto: University of Toronto Press.

Noddings, N. (2002). *Educating moral people: A caring alternative to character education*. New York: Teachers College Press.

O'Sullivan, E. (2003). Bringing a perspective of transformative learning to globalized consumption. *International Journal of Consumer Studies, 27*(4), 326–330. https://doi.org/10.1046/j.1470-6431.2003.00327.x.

Ontario Ministry of Education. (2007). *Ready, set, green! Tips, techniques, and resources from Ontario educators*. Toronto: Queens Printer for Ontario.

Ontario Ministry of Education. (2009). *Acting today, shaping tomorrow: A policy framework for environmental education in Ontario schools*. Toronto: Queens Printer for Ontario.

Ontario Ministry of Education. (2011). *The Ontario curriculum grades 1–8 and kindergarten programs. Environmental education: Scope and sequence of expectations*. Toronto: Queens Printer for Ontario.

Ontario Working Group on Environmental Education. (2007). *Shaping our schools, shaping our future: Environmental education in Ontario schools*. Toronto: Queens Printer for Ontario.

Passmore, J. H. (1972). *Outdoor education in Canada: An overview of current developments in outdoor education and environmental studies*. Toronto: Canadian Education Association.

Pedretti, E., & Nazir, J. (2014). Tensions and opportunities: A baseline study of teachers' views of environmental and outdoor education. *International Journal of Environmental and Science Education, 9*(3), 265–283. 10.12973/ijese.2014.215a.

Polkinghorne, D. E. (1989). Phenomenological research methods. In R. S. Valle & S. Halling (Eds.), *Existential–phenomenological perspectives in psychology* (pp. 41–60). New York: Plenum.

Rickinson, M. (2001). Learners and learning in environmental education: A critical review of evidence. *Environmental Education Research, 7*(3), 207–320. https://doi.org/10.1080/13504620120065230.

Russell, C., Bell, A. C., & Fawcett, L. (2000). Navigating the waters of Canadian environmental education. In T. Goldstein & D. Selby (Eds.), *Weaving connections: Educating for peace, social and environmental justice* (pp. 196–217). Toronto: Sumach.

Santopietro, G. D. (1995). Raising environmental consciousness versus creating economic incentives as alternative policies for environmental protection. *Journal of Economic Issues, 29*(2), 517–524. https://doi.org/10.1080/00213624.1995.11505688.

Sauvé, L. (2005). Currents in environmental education: Mapping a complex an evolving pedagogical field. *Canadian Journal of Environmental Education, 10*(1), 11–37.
Schlegelmilch, B. B., Bohlen, G. M., & Diamantopoulos, A. (1996). The link between green purchasing decisions and measures of environmental consciousness. *European Journal of Marketing, 30*(5), 35–55. https://doi.org/10.1108/03905696101118740.
Stevenson, R. B., Brody, M., Dillon, J., & Wals, A. E. J. (Eds.). (2013). *International handbook of research on environmental education*. New York: AERA Routledge.
Wals, A., & Dillon, J. (2013). Conventional and emerging learning theories: Implications and choices for educational researchers with a planetary consciousness. In R. B. Stevenson, M. Brody, J. Dillon, & A. E. J. Wals (Eds.), *International handbook of research on environmental education* (pp. 253–661). New York: AERA Routledge.

Chapter 8
Indonesian Adventures: Developing an Ecology of Place on Sulawesi Utara

Vajiramalie Perera, Wiske Rotinsulu, John Tasirin, and David Zandvliet

Abstract In this chapter, we develop an island metaphor to communicate our emerging framework for EE in Indonesia. The concept of an island is a powerful metaphor in everyday speech as well as in the disciplines, and we use it here as an attempt to clarify our meaning of community. Beyond the metaphor, islands have also played a major role in the realm of knowledge construction (e.g., descriptions of isolated gene pools were seen as instrumental in the development of Darwinism, and these processes were described as taking place within the "Malay" archipelago by Wallace). Social anthropology also uses islands implicitly in the description of isolation and boundedness in cultural systems. In the case of Sulawesi Utara, this insularity is a strong descriptive metaphor but also describes an ecological reality for this region. Environmental learning then can draw on the functions, intersections, and relations of place-based education. Our discussion is informed by a place-based island metaphor for ecological education that emerged from the development work for a field school conducted in Indonesia. Place-based education (in our view) discards a one-sided view of education by taking as its first assumption that education is both "about" and "for" defined communities. This perspective then informed a study of place-based education on Sulawesi Utara and the design of the field school for teachers that we conduct there.

8.1 Prologue

> From a look at a globe or a map of the Eastern hemisphere, we shall perceive between Asia and Australia a number of large and small islands forming a connected group distinct from those great masses of land, and having little connection with either of them. Situated upon the Equator, and bathed by the tepid water of the great tropical oceans, this region enjoys a

V. Perera (✉) • D. Zandvliet
Simon Fraser University, 5886 Coquitlam St, Chilliwack, BC V2R 0A1, Canada
e-mail: vajiramalie@gmail.com; dbz@sfu.cal

W. Rotinsulu • J. Tasirin
Universitas Sam Ratulangi, Kampus Unsrat, Bahu-Manado 95115, Indonesia
e-mail: wiske_rotinsulu@yahoo.com; jtasirin@gmail.com

G. Reis, J. Scott (eds.), *International Perspectives on the Theory and Practice of Environmental Education: A Reader*, Environmental Discourses in Science Education 3, https://doi.org/10.1007/978-3-319-67732-3_8

> climate more uniformly hot and moist than almost any other part of the globe, and teems with natural productions which are elsewhere unknown. The richest of fruits and the most precious of spices are Indigenous here. (Alfred Russell Wallace 1890, p. 11)

At its initial conception, it was envisioned that an elective field school program in Indonesia (as part of the undergraduate minor in environmental education (EE)) would appeal to teacher candidates who were interested in pursuing a career in this domain. The focus of the field school was to be on the development of students' environmental perspectives through an investigation of diverse ecological contexts in an international setting. Most importantly, we believed that the adventurous aspect of the field school would create a unique place-based learning opportunity for students.

Indeed, Indonesia offers students a rich opportunity to explore intercultural and international approaches in education as part of a unique semester-abroad experience. The program includes two courses: environmental education (104 h) and directed studies in environmental education (52 h). These are delivered in cooperation with local partners and involve day excursions or multiple day-trips to local communities, national parks, meetings with community organizations, exchanges with university students, cultural events, presentations by university representatives, and volunteer work with community organizations and schools. These experiences form the core of a place-based curriculum centered in various island communities on Sulawesi Utara, located on the northern tip of the island formerly known as Celebes. Our interactions with the community there set up a dialectic that leads to service learning opportunities and formulation of self-directed studies in support of local community. In this, students were encouraged to develop a project that supported local efforts at environmental remediation, communication, or curriculum development.

The course follows in the footsteps of explorer and naturalist Alfred Russel Wallace, who traveled extensively throughout the region in the mid-eighteenth century and codeveloped his theory of natural selection with Darwin. The place also witnessed waves of intrepid Dutch traders prior to that, who colonized much of modern-day Indonesia, centralizing governance in Jakarta and developing its botanical treasures into a three-century monopoly on the trading of coveted spices such as cloves and nutmeg. In this context, the study program for the field school provides opportunities for students to consider environmental perspectives in both rural and urban settings while also considering theories associated with critical and place-based education such as those of decolonization and re-inhabitation as described by David Grunewald (2003).

8.2 "We're Going on an Indonesian Adventure"

This chapter also researches and describes through an ethnographic perspective a core EE course taught in the international setting of Indonesia. The course examines in detail the educational problems entailed in developing human awareness and understanding of the environment through a dialectic of the global and local. It also explores environmental issues through multidisciplinary lenses which include scientific inquiry, aesthetic, and social justice perspectives to a consideration environmental ethics. It also relates historical and contemporary issues in human–environment interactions to school curricula from the elementary to the secondary level. Designed for prospective teachers, the course is a core for the minor in environmental education at Simon Fraser University (SFU).

> From volcano geology to highland agriculture studies to coral reef ecology, students can look forward to a range of coursework about the environment. With the current trends in sustainable development, the main objectives of this field school will be on developing students' environmental perspectives through hands on fieldwork in an international setting. Also, the program provides an introduction to how educators can promote environmental and ecological thinking across all grade levels and all subject areas. (Excerpt from the 2011 brochure)

This study describes the development and delivery of an international field school in Indonesia that has been offered since 2007. The field school represents collaboration with the Universitas Sam Ratulangi in Manado that extends the university's earlier capacity building in the region with the CIDA-funded Eastern Indonesian Universities Development Project (EIUDP) through the 1990s (Munro and Dagg 1998). We employ an auto-ethnographic reporting style where "researchers constitute their own object of research so that the knowing subject and the research object become one" (Roth 2005, p. 109). In addition, we explore the "lived experience" (van Manen 1990) of students and faculty participating in the program and relate to our evolving conceptions of a critical place-based education in situ and the relationship of this experience to various learning outcomes associated with environmental and ecological education.

The model for ecological education discussed in this chapter builds on existing theories of place-based education (Smith 2002) while describing a deepening partnership between Indonesian and Canadian universities in the offering of an international field school. It draws on participatory action research methods (Gaventa 1988) to relate the perspectives of various stakeholders including community members, students, teachers, and university faculty throughout our work.

The goal for place-based education of this nature is to describe an ecology that emphasizes the embeddedness of human societies and cultures within localized places or communities (Bowers 1999). This conceptual model describes a range of ecological, sociocultural, and technical influences that frame educators' interpretations of curriculum. Our field school explores this model within the context of island communities where we apply the framework in specific ways – contributing to the development of future educators through their field work, community service, and curriculum development (Fig. 8.1).

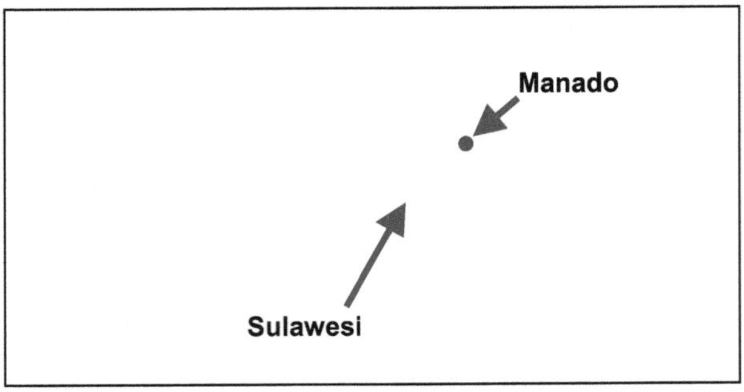

Fig. 8.1 Location of the Indonesia Field School (Adapted from: https://www.cia.gov/library/publications/the-world-factbook/geos/id.html)

8.3 "Island" as Metaphor and as Place

> The position of Celebes (Sulawesi) is the most central in the Archipelago. Immediately to the north are the Philippine islands; on the west is Borneo; on the east are the Molucca islands; and on the south is the Timor group – and it is on all sides so connected with these islands by its own satellites, by small islets, and by coral reefs, that neither by inspection on the map nor by actual observation around its coast, is it possible to determine accurately which should be grouped with it, and which with the surrounding districts. (Russel Wallace 1890, p. 380)

The Indonesia Field School uses an island metaphor to communicate a developing framework for place-based education that is suitable for an archipelago nation like Indonesia. The concept of an island is a powerful metaphor in everyday speech as well in disciplines such as sociology or science, and we use it here as an attempt to clarify our meaning of community. Social anthropology also uses islands implicitly in the description of isolation and boundedness in cultural systems (Ericksen 1993). In the case of Sulawesi Utara, this insularity is a descriptive metaphor that also depicts the ecological reality for the region as a distinct cultural region and a biodiversity hotspot with high levels of endemism.

Islands have also played a major role in the realm of knowledge construction in science: descriptions of isolated gene pools, for example. Indeed, Wallace (1890) first described a mechanism for natural selection in the Malay Archipelago:

> Isolation is no doubt an important aid to natural selection, as shown by the fact that islands so often present a number of peculiar species … an examination of its animal productions shows Celebes (Sulawesi) to be at once the poorest in the number of its species, and the most isolated in the character of its productions, of all the great islands in the Archipelago … A considerable number of its animal forms are so remarkable, as to find no close allies in any part of the world …. (Russel Wallace 1890, p. 31)

Today, nearly 70% of all species on the island of Sulawesi are still considered endemic to the island (John Tasirin, personal communication). These include a range of endangered mammals, such as the anoa (a dwarf species of water buffalo), babirusa (the so-called pig deer), bear, cuscus, maleo bird, and a number of rare species of primate (i.e., the black crested macaque and spectral tarsier). This high degree of endemism is attributed by scientists – Darwin and Wallace included – to the isolated physical location of these islands and the extremely active seismic

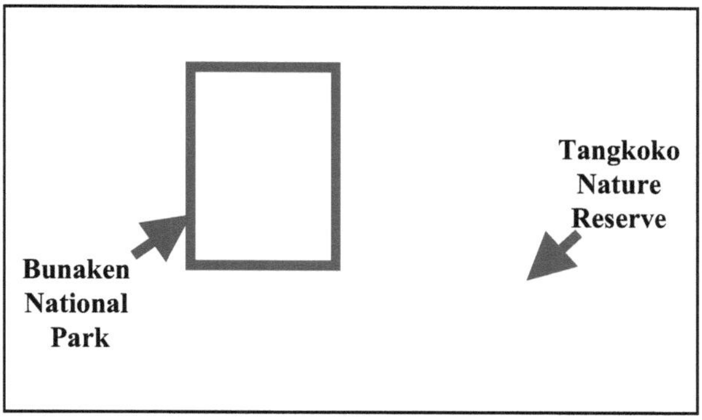

Fig. 8.2 Manado and the Minahasa region (Adapted from: http://www.minahasa.net/en/about-map.html)

nature of the surrounding region. It is a place where the Asian and Australian tectonic plates collided and brings about a blend and evolution of floral and faunal species from the Oriental and Australasian bioregions of the world.

Geographically, islands can be said to be mountains that emerge from the bottom of the sea to tower above the water. In the case of Sulawesi Utara, this is literally the case as active and dormant volcanoes, fringed by coral reefs, form the backdrop for various forms of community life. The structural similarity of the "phenomenological order" between such types of landscape as an island in the middle of the open sea allows for the similarity of metaphorical meanings for an island or mountain (Lehari 2005). In this, the precondition of an island's metaphorization is its existence in our own environmental experience. Put more simply, "an island is not an island, until you go there" (Zandvliet and Brown 2006). That is, the isolating component of this methaphorization of island can only be described by those "on island" looking back out rather than those viewing the phenomenon from other vantage points (Fig. 8.2).

8.3.1 Sulawesi Utara

> The first night we were there, a bunch of us caught our first Indonesian sunset. I remember us being in awe at the colours of the sky, the Manado Tua volcano backdrop and the calmness of the Celebes Sea. (Vajiramalie/RA)

Locating a field school in the Minahasa region of Sulawesi was an ideal design decision for modeling our concept of place-based education owing to both its cultural and biological diversities. The region must be truly experienced to be understood:

> The trees are more covered with lichens and mosses, and the ferns and tree–ferns are finer and more luxuriant than I had been accustomed to seeing ... The forest, however, is most luxuriant. Noble palms, Pandani, and tree–ferns are abundant in it, while the forest trees are completely festooned with Orchideae, Bromeliae, Araceae, Lycopodiums, and mosses … Ordinary stemless ferns abound; some with gigantic fronds ten or twelve feet long, others barely an inch high; some with entire and massive leaves, others elegantly waving their finely–cut foliage, and adding endless variety and interest to the forest paths …. (Russel Wallace 1890, p. 349)

The reasons for this abundance are clear: first, the region has been recognized as a biodiversity hotspot by scientists (UNESCO 2005) and is located well within the famed coral triangle. Second, located on the northern tip of an island, Sulawesi is itself surrounded by other small volcanic island groups to the north and south and is bathed by the warm currents of the Celebes Sea.

8.4 Learning in Place

> Being from (an island) not so far from Indonesia, I was quite used to the climate and living conditions. I enjoyed the hot weather, spicy food and beautiful beaches. Almost every facet of Indonesia reminded me of home … Still I was a little afraid that I would not get experience of being outside my comfort zone … little did I know that the experiences that have changed my life came from our group and the ordeals that we had to endure together …. (Vajiramalie/RA)

The learning activities designed for the field school allowed students to build community through student dialogue while also experiencing diverse natural settings such as the active volcanoes of *Lokon* and *Mahawu* in the highlands of Tangkoko Nature Reserve (a protected lowland forest) and *Pulau Bunaken* (with its fringing coral reefs). These activities first exhibited the adventure aspect of our experience perhaps challenging students physically and/or mentally while also providing a firsthand look at the extraordinary diversity of ecosystems present in the Minahasa region.

Our natural history adventures lasted from short trips of a few days to longer trips of a week or more, as we explored the surrounding environment from a geological and/or biological perspective. These scientific experiences were complemented with visits to cultural sites such as Pasar tradisional (the traditional market) in *Tomohon* or *Airmadidi* (with its ancient Minahasan burial crypts known as Waruga). Many other activities explored traditional local (sustainable) technologies, such as bamboo cooking, fish traps, and other forms of traditional knowledge that have been part of these local cultures for thousands of years.

> My fondest memory in Indonesia is definitely the outing to Tangkoko National Park. The army of macaques that surrounded me that afternoon was unreal. All it took was one alarm call from one macaque and a crowd of them appeared, it was beyond what I expected to see on the field school … it was an amazing experience…. (Steven, field school participant) (Fig. 8.3)

Important to our goal of modeling ecological education practices: several different indigenous groups were present in the Minahasa highlands along with an intermingling of migrant populations from neighboring islands in the coastal region. It is a textbook case of cultural boundedness that has been eased by the migration of other groups to Minahasa. In the highlands, nine tribes have merged to form the dominant Minahasan cultural group. Much of this intermingling was accelerated by the colonial presence of early Portuguese or Dutch traders. While Portuguese were in the region only briefly, the Dutch planted coffee or spices leaving a lasting cultural impact. This included the establishment of Christianity as the dominant religion in the region and the persistence of the Dutch language among elders in the community.

Fig. 8.3 Black crested macaque (Tangkoko), Waruga complex (Airmadidi), and bamboo cooking (Tomohon) (© D. Zandvliet)

8.5 Dutch Influence

> The country was divided into districts, and the system of Controlleurs, which had worked so well in Java, was introduced. The Controlleur was a European, or a native of European blood, who was the general superintendent of the cultivation of the district, the adviser of the chiefs, the protector of the people, and the means of communication between both and the European Government ... under the direction of the Controlleurs most of the houses were rebuilt on a neat and uniform plan. It was this interesting district which I was now about to visit .... (Russel Wallace 1890, p. 344)

For us, environmental learning in a place like Indonesia draws on the functions, intersections, and relations of a critical form of place-based education. For example, in our "spice of life" assignment, we analyzed the ongoing Dutch influence in the region as a historic spice-producing region for the Dutch East India Company (or VOC) through a variety of activities. In this assignment, we describe differences between colonial relations to place and other (indigenous) relations to land, where a traditional ecological knowledge of the land is integral to "being and knowing in the world" (Meyer 2008, p. 211), this as seen through the globalizing practice of spice trading (Fig. 8.4).

> The major source of new and useful spices turned out to be in South East Asia where spices such as pepper, cardamom, ginger, cinnamon, cloves and nutmeg were indigenous. After a brief period of Portuguese dominance in the region, Dutch traders drove all other European nations out of the region and gained a virtual monopoly of the trade in nutmeg and cloves ... Dutch influence grew with the increasing wealth of the enterprising Dutch East India Company until the British and French smuggled out plants and began competing plantations in other regions. The Dutch East India Company eventually collapsed in 1799 .... (Excerpt from Spice of Life assignment guidelines)

This colonial relation to place in Minahasa is still evident (alongside indigenous perspectives) as large clove and coffee plantations persist in and around many villages such as Rurukan. Despite the end to Dutch colonization in the 1950s, this relationship remains a complex one as many Minahasans fought alongside the Dutch (against the Javanese) during the Indonesian war of Independence. Some Minahasan elders continue to believe that they enjoyed greater economic and

Fig. 8.4 Nutmeg, coffee and cloves, profitable Indonesian commodities in the Dutch colonial time (© D. Zandvliet)

political autonomy under colonial Dutch rule than under their current leaders in Jakarta (personal communication), this, despite the recent efforts by the central government in Jakarta to grant greater autonomy in the region.

8.6 Kawanua

> In the Minahasa language the word Kawanua often means inhabitant of the country, or wanua–wanua (people) who are one, or Mina–Esa (Minahasa People) …. (John Tasirin, local lecturer, personal communication)

The meaning of the word Kawanua is thought to be derived from the earlier proto-Malay word Wanua, meaning a dwelling or domain, and later it developed to mean village, negeri (country), or nation. At the same time, in Minahasan culture, the word Wanua had come to take the deeper (ecological) meaning of country or village.

For the first part of our stay in Indonesia, we typically attempt to embody the concept of Kawanua as we try to inhabit the area: imbedding ourselves in typical highland patterns of daily life while observing/participating in the rituals which surround us in this largely subsistence-based agricultural region. These patterns include (in part) communal agricultural practices, local or small cottage industries, and intergenerational vocational education programs.

> We settled into the nice cool weather and fresh mountain air… we were invited to a "wedding breakfast dinner" at a close–by village. They had interesting customs and were very accommodating to us …. (Veronica, field school participant)

The techniques we use to imbed ourselves in community life include a process of community mapping that involves open-ended, ecological inquiry into the life of a village or locality. As part of this process, students are asked to observe, analyze, and represent their personal experiences of community to each other through each of three lenses. The first (sociosphere) involves people-watching and otherwise dialoging and informally interacting with locals (using interpreters where necessary). The second (technosphere) involves investigating the how and why things are done

Fig. 8.5 Tree ferns, farm collaborative, market produce, and rice paddy all representing the community (© D. Zandvliet)

a certain way (e.g., building or farming techniques). Finally, the lens of ecosphere considers the all-encompassing natural environment where all of these other activities take place (Fig. 8.5).

> We were sent on our community–mapping project … I loved this part of the field school. Walking through the village and talking to random strangers was quite exciting. I really enjoyed walking through paddy fields and visiting the pagoda …. (Jessica, field school participant)

8.7 A Critical Approach to Place-Based Education

On one occasion, shortly after the community mapping activity and just after a visit to the traditional market in Tomohon, one student expressed a deep concern: he noted that many endangered species we had observed on excursions to natural areas such as the Tangkoko Nature Reserve (i.e., rats, bats, snakes, monkeys) had found their way to the market as <u>food</u>. To this point one of the local lecturers (John Tasirin) responded:

> This is the problem we are trying to address. In the Minahasa region, we have one of the highest levels of biodiversity on the planet … but we have a culture that eats everything...

Later in the field school, this same lecturer related that the actual relationship might be a bit more complex than originally stated. During a personal communication, he indicated that this cultural reality underscores a need for critical forms of environmental education as the loss of an endangered species such as *Macaca nigra* (the black crested macaque) would also see a loss of significant cultural capital. However, many Minahasans (himself included) still consume local (indigenous) plants or animals as a kind of cultural practice – often linked to their personal identity as Minahasans in an increasingly globalized and interconnected society (Fig. 8.6).

The notion of place-based education has been well described by David Orr (1992) and David Sobel (1993), while related ideas have been expanded on by others (e.g., Woodhouse and Knapp 2000). The idea of place-based learning connects theories of

Fig. 8.6 Traditional highland food, bajo with fish trap, UNSRAT biologist, John Tasirin (© D. Zandvliet)

experiential learning, contextual learning, problem-based learning, constructivism, outdoor education, indigenous education, and environmental education. The above incident underscores the need for critical approaches in the study of place-based education in that it has both ecological and cultural dimensions. In support of what he described as a critical pedagogy of place, David Gruenewald (2003) states that our educational concern for local space (community in the broad sense) is sometimes overshadowed by both the discourse of accountability and by the discourse of economic competitiveness to which it is linked. Echoes of this idea can be seen in the increasing influences of globalization that are observed in many urban areas of Indonesia, but also in the Dutch colonial influences that are still visible in the Minahasa region. As this case illustrates, the consumption of local foods becomes commodified at a local market rather than a home-based cultural practice. This, in turn, contributes to the growing conservation issues as they are influenced by increasing urbanization caused by the fact that more and more Indonesians leave their rural communities for jobs in the larger cities.

8.8 Menado (Manado) and the Coastal Region

> The little town of Menado is one of the prettiest in the East. It has the appearance of a large garden containing rows of rustic villas with broad paths between, forming streets generally at right angles with each other. Good roads branch off in several directions towards the interior, with a succession of pretty cottages, neat gardens, and thriving plantations, interspersed with wildernesses of fruit trees. To the west and south the country is mountainous, with groups of fine volcanic peaks 6,000 or 7,000 feet high, forming grand and picturesque backgrounds to the landscape. (Russel Wallace 1890, p. 341)

The Manado that Wallace described some 150 years ago was a product of coastal human migration patterns to this productive and scenic region. However, these were magnified by increased development during Dutch colonial times. Today, the region has seen more dramatic growth as a global tourism destination (due to its proximity to the Bunaken Marine Park) and as an important regional center for government, agriculture, mining, and commerce. It increasingly resembles any large city with popular shopping complexes, fast food, chic hotels, and trendy restaurants. As the city grows, so do its problems: urban encroachment on protected areas, solid waste (i.e., plastic migrating to coral reefs), air pollution, and traffic congestion. In short, Manado faces many of the same sustainability issues that larger urban areas face, for example, how to balance development without compromising the important ecological and cultural capital present in the surrounding regions.

This then was to be the true urban adventure as an important program activity is the act of living in and experiencing the city of Manado. We do this by shopping or eating, by interacting with school children while teaching English at local schools, by meeting with local NGOs, and by attending lectures at the local campus of Universitas Sam Ratulangi. The content of these activities provides participants with a host of information on topics such as marine ecology and the natural history of Sulawesi Utara, but they also include topics of a more urban focus including issues of language and culture, public health, education, and a variety of other development-related issues. The intent of this is to allow students to draw connections between their everyday lives and the issues apparent in the surrounding physical/cultural environments that support their (largely) urban lives.

Overall, we believe that the inclusion of a form of urban EE in our field school program gives students a greater appreciation of the diversity of cultural and environmental systems around them as well as an urge to act to overcome existing problems. Indeed, it is a synergy of universities, schools, and community-based organizations that are together struggling to promote more sustainable living here in Sulawesi Utara and abroad. Still, common urban practices such as hyper-consumerism, environmental depletion, and socioeconomic inequality also need to be seen as contributing to the range of environmental problems. All educators, as individuals and professionals, are expected to respond to the collective demand for a greener socially just world but are themselves implicated through their everyday lives:

> My experience has allowed me to develop another set of lens in viewing our home in Canada. It has made me more aware of how my value system, decisions, and actions all influence my immediate surroundings and may encompass environmental consequences.... (Jessica, field school participant)

8.9 Island Reflections

Bunaken National Park, only a few watery kilometers from the growing city of Manado, encompasses a coral ecosystem and has been recently proposed as a UNESCO World Heritage listed site. It is also considered by scientists to be near the epicenter of marine biodiversity on the planet (UNESCO 2005). This location marks a final destination for course activities. Here students finalize coursework and reflect deeply on their learning throughout the program. Portfolio presentations and action projects comprise the bulk of the activity while on island, and these are

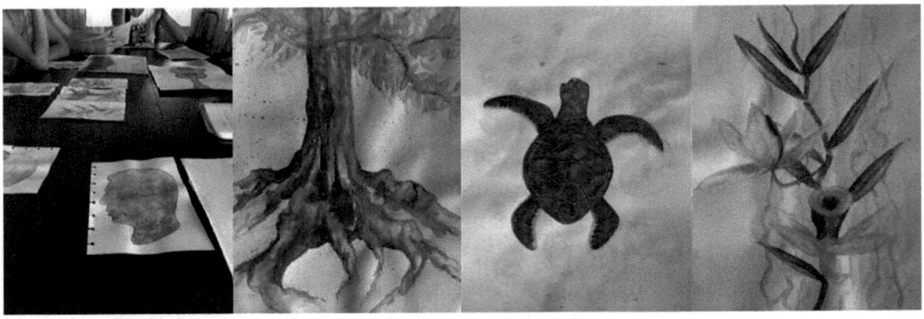

Fig. 8.7 Artistic expression (water colors) as an example of the portfolio process by Merissa Robson (© D. Zandvliet)

supplemented by a number of field excursions to beaches, reefs, villages, schools, and other excursions (i.e., dolphin or whale-watching) – all of these occurring within the park boundaries. Students engage in peer-teaching activities where they also share knowledge gained about the local ecology and culture and try out ideas or techniques related to the practices or theories of environmental education (Fig. 8.7).

> During the last module of the course (in the form of a retreat), you will share artifacts, ideas, writings and lessons using a portfolio of your own design, to demonstrate learning and connections made between conceptual frameworks for Environmental Education and your own developing practice .... (Excerpt from 2014 course syllabus)

Through deep reflection within the portfolio process, participants come to understand that place-based and experiential programs can examine the complexity of natural systems. Human interaction with/in these systems is also examined while they explore how holistic forms of education can help to develop a sense of respect and appreciation for the natural world. Aesthetic appreciation, along with a scientific understanding of nature, encourages them to act to protect and sustain the environment. The actions they intend will take the form of detailed directed studies proposals that students compose in order to complete their studies in the semester-long program.

Student-designed (directed studies) projects take the form of curriculum writing projects, community action projects, or various other types of creative works (McClaren and Hammond 2005). Throughout, the principles for organizing and conceptualizing EE are demonstrated and applied by students. These aspects include a consideration of complex systems, aesthetics, environmental responsibility, and the practice of an environmental ethic (BC Ministry of Education 2007). As educators, we need to facilitate understandings of what constitutes responsible environmental action but also to help students to think/act responsibly in both their personal/professional lives.

8.10 Ecological Education

Ecological frameworks (such as the one described here) apply the principles of ecology, derived from the Greek root "oikos" (household), to an examination of the relationship of all living things with their environments and with one another as living and interdependent systems (Bowers 1999). So too, in a philosophical sense, ecological notions such as community or *Kawanua* apply to our conception of the human–world relationship and to the theory and practices of education. Ecological frameworks also aim to build on a specific understanding: that humankind is an interconnected part of both the human and natural worlds (Smith 2002). Further, to understand ecologically is to make sense of the human world as part of, not apart from, nature; it is to understand humankind's "interconnectivity in life" (Grunewald 2003, p. 6).

> It was in North Sulawesi that I found a true appreciation and realization as to just how interconnected our environment is ... Living here (in Canada) I feel disconnected from my immediate environment. In Indonesia, I knew where the banana I was eating came from as I saw the (Minahasa) boy cut it down from a tree, and I knew where my fish was from as I saw it swimming in the ocean .... (Amanda, field school participant)

Understanding ecologically also has an emotional core: one's knowledge about ecological processes and principles is meaningful due to an emotional attachment to the world and all of its living communities. According to the authors, the consideration of an inclusive, ecological framework for education at once responds to a critique of the mainstream organization of curricula by providing for alternative issue-based and place-based pedagogies while allowing teachers to interpret curriculum in ways that refocus learning on and in communities. Teaching within an ecological framework focuses energies on the importance of quality of life within communities while assisting students in the development of a sense of place within them. While others such as Gruenewald (2003) and Smith (2002) make arguments for place-based or community-based models of learning, we attempt to take this view further by also describing the need for critical and embodied approaches in their implementation. Central to this is the idea that ecological frameworks for education recognize that our assumptions about effective teaching are best enacted when actions are deeply embedded within the complexity and authenticity of real communities:

> Teaching in Indonesia at both a public and private school made me realize two things. First, just how much I love working with children. Second, that providing people with a proper education can happen anywhere. You don't have to have big fancy schools, projectors, smart boards etc. what is essential is having capable and caring educators. No matter where I go in life, whether in the traditional classroom or in a government office, I aspire to be an educator. (Amanda, field school participant)

> By the end of my two months in North Sulawesi, Indonesia "Terima kasih" (thank you) rolled off of my tongue with no trouble. It is a phrase I hope I never forget as I am forever thankful for my time and experience in Indonesia ... (Rachel, field school participant)

8.11 Final Thoughts

> For me, Indonesia is like that door that 'Alice in the Wonderland' enters. Indonesia was an experience of many 'firsts' for me such as swimming in the ocean, visiting a black sand beach, hiking ... just to name a few .... (Jessica, field school participant)

As described here, the idea of place-based learning connects theories of experiential learning, contextual learning, problem-based learning, constructivism, outdoor education, indigenous education, and EE. However, in describing the need for a critical pedagogy of place, Gruenewald (2003) writes that our educational concern for local space (community) can be overshadowed by the discourse of accountability and economic competitiveness to which it is linked. This assertion has also been witnessed through our experiences in Sulawesi Utara as the influence of globalization is increasingly felt within the local communities we visit during the field school. Place becomes a critical construct to its opponents, not because it is in opposition to economic well-being but because it challenges assumptions about the dominant metaphor of progress and its embedded values.

An ecological framework for EE breaks from this mold by taking as its first assumption that education is both about and for community. Ecological conceptions of education should place an emphasis on the inescapable embeddedness of humans and their technologies in natural systems. Instead of conceptualizing nature as *Other*, it involves the practice of viewing humans as one part of the complex natural world, where human societies and cultures are a product of the interactions between our species and the places in which we find ourselves (Smith 2002). Such an approach also negates issues of right or wrong and allows individuals or groups to consider multiple (cultural) perspectives on an issue or problem. This allows for sociocultural critiques to be placed alongside the usual scientific or economic considerations. The implications of this shift in thinking could be profound for the future of environmental education:

> There is an unexplainable distinction between departing and arriving – so much fills the space between the two and it seems as if words cannot convey the weight of the experience. I am here, on the plane ... about to take off from the Manado airstrip. We are ... full of memories, experiences, new destinations, sadness, future plans, and we have shared something that is more than we ever envisioned it to be ... Never before have I been so ignited with future aspirations and felt the possibility of it all and how anyone, anywhere can impact who we are and how we perceive the world. (Veronica, field school participant)

As a final reflection, the Indonesia Field School director (David Zandvliet) also attributes the implementation of this field school as a transformative moment in his own professional career:

> In planning the logistics and theory to be implemented in this program, I had the opportunity to travel extensively in Indonesia and visited many places in Sulawesi including Tomahon, Tangkoko, Manado and Bunaken ... all of these are now special places in my own experience and each place teaches our students a great deal about the interactions of culture and environment. Despite this, the experience of running the field schools in this region has changed me in ways that I will never be able to fully understand, much less describe ...

Thinking through our experiences in Indonesia has caused us to view the concept of home quite differently and the purposes of education too. So, is ecological education the answer to this form of economic inequality so prevalent in regions like Indonesia? Many theorists argue that a place-based or "ecological education" may help to alleviate this condition. This assertion has also been strengthened through our experiences in Sulawesi as the influence of globalization is increasingly felt within the local communities visited. Place becomes a critical construct to its opponents, not because it is in opposition to economic well-being but because it challenges assumptions about the dominant metaphor of progress and its embedded values. I believe that in summary, this is an important consideration for all those interested in the internationalization of education, as the diverse cultural practices we witness in other locations can function to "make the familiar, strange" in the places we now call home.

Questions

Discussion questions that could help instructors and students to engage in meaningful conversation about the ideas presented in this chapter.

1. List the differences and similarities of the terms "place-based education" and "environmental education" based on their description in this chapter. Are they compatible? What would be the benefit(s) of meshing the two together in formal, non-formal or informal learning environments?
2. How do aspects of culture and environment interact in this description of a field school?
3. How would you describe the meaning of "kawanua" in this study, and can you think of similar ideas developed in other settings?

References

Bowers, C. (1999). Changing the dominant cultural perspective in education. In G. A. Smith & D. R. Williams (Eds.), *Ecological education in action: On weaving education, culture and the environment* (pp. 161–178). Albany: State University of New York Press.

British Columbia Ministry of Education. (2007). Environmental learning and experience: An interdisciplinary guide for educators. Retrieved from http://www.bced.gov.bc.ca/environment_ed/.

Ericksen, T. E. (1993). Do cultural islands exist? *Social Anthropology, 1*(1), 133–147. https://doi.org/10.1111/j.1469-8676.1993.tb00246.x.

Gaventa, J. (1988). Participatory research in North America. *Convergence, 21*(2/3), 19–27.

Gruenewald, D. (2003). The best of both worlds: A critical pedagogy of place. *Educational Researcher, 32*(4), 3–12.

Lehari, K. (2005). Island. Retrieved from http://www.eki.ee/km/place/pl03/Place3_Lehari.pdf.

McClaren, M., & Hammond, B. (2005). Integrating education and action in environmental education. In E. Johnson & M. Mappin (Eds.), *Environmental education and advocacy: Changing perspectives of ecology and education* (pp. 267–291). New York: Cambridge.

Meyer, M. A. (2008). Indigenous and authentic: Hawaiian epistemology and the triangulation of meaning. In N. K. Denzin, Y. S. Lincoln, & L. T. Smith (Eds.), *Handbook of critical and indigenous methodologies* (pp. 217–232). Los Angeles: Sage.

Munro, J., & Dagg, C. (1998). *Canadian and Indonesian universities work towards sustainable development in eastern Indonesia*. University Leaders for a Sustainable Future. Publications, 2(2). Retrieved from http://www.ulsf.org/publications_declaration_index.html

Orr, D. (1992). *Ecological literacy*. Albany: State University of New York Press.

Roth, W. M. (Ed.). (2005). *Auto/biography and auto/ethnography: Praxis of research method*. Rotterdam: Sense Publishers.

Smith, G. A. (2002). Place-based education: Learning to be where we are. *Phi Delta Kappan, 83*(8), 584–594.

Sobel, D. (1993). *Children's special places*. Tucson: Zephyr Press.

UNESCO. (2005). World heritage list. Retrieved from http://whc.unesco.org.

van Manen, M. (1990). *Researching lived experience: Human science for an action sensitive pedagogy*. New York: State University of New York Press.

Wallace, A. R. (1890). *The Malay Archipelago: The land of the orang–utan, and the bird of paradise. A narrative of travel, with sketches of man and nature* (10th ed.). London: Macmillan.

Woodhouse, J., & Knapp, C. (2000). *Place–based curriculum and instruction: Outdoor and environmental education approaches*. Charleston: ERIC Clearinghouse on Rural Education and Small Schools.

Zandvliet, D. B., & Brown, D. R. (2006). Framing experience on Haida Gwaii: An ecological model for environmental education. *Canadian Journal for Environmental Education, 11*(1), 207–219.

Part III
Environmental Education in the Context of Schools

Chapter 9
Crisis and Recovery in Environmental Education: The Case of Greece

Constantinos Yanniris and Myrto Kalliopi Garis

Abstract The challenges posed by the global economic slowdown are being felt by environmental education (EE) providers worldwide. In Greece, the national framework of EE was severely affected by the localized effects of the Eurozone debt crisis. Notwithstanding the loss in materials, school projects and educational programmes, this catastrophic event offers a rare window of opportunity to study and improve the systemic strength of EE infrastructure. This chapter explores the institutional, structural and pedagogical characteristics that contribute to the resilience of EE frameworks. Resilient systems are more likely to survive future socio-economic perturbations in a world of growing uncertainty.

During the last three decades, a number of countries have introduced environmental education (EE) in their national contexts (Palmer 1998). As international guidelines were absorbed by diverse educational systems, a multitude of approaches appeared (Sauvé 2005). This chapter examines the case of Greece, where an institutionalized, school-based version of outdoor EE has taken effect since the early 1990s (Katsakiori et al. 2008). The focus is on the effects of the global financial crisis on local EE frameworks and infrastructure in the country. The Greek experience may prove didactic for countries that seek to enhance the resilience of their own systems.

In Greece, EE projects are implemented at the school level by environmental teams that typically involve one or two teachers and a group of 20–25 students. These environmental teams are formed on a voluntary or elective basis and require a year-long commitment (Tsaliki 1998). In each team, the topic of their study is

C. Yanniris (✉)
Department of Integrated Studies in Education, McGill University,
3700 McTavish Street, Montréal H3A 1Y2, Canada
e-mail: constantinos.yanniris@mail.mcgill.ca

M.K. Garis
Faculty of Environmental Planning and Design, Université de Montréal,
2940 Côte–Sainte–Catherine, Montréal H3C 3J7, Canada
e-mail: myrto.gari@umontreal.ca

G. Reis, J. Scott (eds.), *International Perspectives on the Theory and Practice of Environmental Education: A Reader*, Environmental Discourses in Science Education 3, https://doi.org/10.1007/978-3-319-67732-3_9

decided by consensus; thematic content spans from biodiversity and resource management to local history, oral traditions and outdoor living (Greek Ministry of Education 2010). This interdisciplinary approach permits and moreover requires the participation of teachers from various backgrounds and disciplines. The EE projects can be realized in various ways, including guided inquiry, interviews and place-based and cooperative learning, but an outdoor component is essential (Georgopoulos 2014). Towards the end of the school year, these environmental teams are expected to present the deliverables of their projects, often in a public event open to the local communities (Mavrikaki et al. 2004).

9.1 Historical Development of Environmental Education in Greece

The first steps of EE in Greece were taken in 1977 when the National Council for Planning and Environment formed an interdisciplinary team that, in tandem with the Council of Europe, UNESCO and the European Commission, started to work on the introduction and development of EE within the Greek educational system. In 1980, the first EE seminar held in Athens was organized under the auspices of the Council of Europe. Following that event, 24 teachers from Greece went on EE training courses in Environment Initiative Centres (CPIE) in France between 1980 and 1983 (Flogaitis and Alexopoulou 1991). At the same time, four Greek schools implemented the first EE projects in the country inspired from a Council of Europe campaign for seashore protection (Gardeli 1983). During the mid-1980s, EE projects started to disseminate within the Greek public school system (Gardeli 1986).

Since the early 1990s, the Greek Ministry of Education enacted legislation which provided administrative and infrastructural means in support of EE within the schooling system. Initially, each of the country's 52 administrative districts acquired two EE counsellors selected from the pool of in-service schoolteachers who have had previous experience with EE. As a result, 104 EE counsellors were appointed to encourage, support and oversee environmental school projects in primary and secondary schools (Greek Ministry of Education 1990). Their actions included the organization of EE seminars for in-service teachers and which required that they co-operated with educational authorities at the local level (Ntarzanou 2008).

The same legislative act allowed the Greek Ministry of Education (1990) to establish Environmental Education Centres (EECs). Local municipalities were encouraged to convert out-of-use peripheral school buildings into EECs. In parallel, the Ministry committed to cover the running costs and provide educational personnel to the newly established centres. Indeed, the country's first Environmental Education Centre was launched in 1993, at the mountain village of Klitoria in Achaia (Katsakiori et al. 2008), and it developed its educational programmes based on local ecological resources, such as the surrounding mountain forest and the adjacent Vouraikos Gorge National Park (Kontaras 2004). Despite its remote location, this

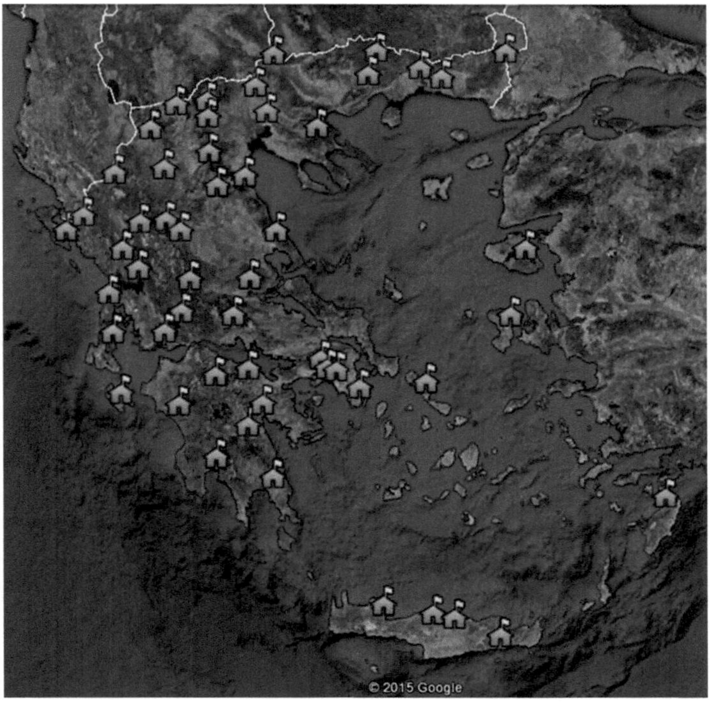

Fig. 9.1 Spatial distribution of the 53 Environmental Education Centres (EECs) operating in Greece (© Yanniris 2016)

centre was soon overwhelmed by an increasing demand from student environmental groups which lined up to visit its facilities (Faragitakis as cited in Kontaras 2004).

In the following years, Environmental Education Centres (EECs) proliferated all over Greece to accommodate a rising number of student environmental groups (Fig. 9.1). A multilateral agreement between local governments, the national government and the European Union secured the accommodation, staff and funding of EECs (Ministry of Education and European Union 2010). Indicatively, between 2009 and 2013, Greek EECs were credited 15,362,258 € from the European Social Fund (Greek Ministry of Education and European Union 2013), which were used to expand their activities and meet their needs of continuing education of the local communities.

Using the ecosystem as an open laboratory, EECs developed interdisciplinary programmes for primary and secondary school students, teachers and adult learners. During the 2014–2015 school year, 471 one- to three-day long educational programmes were offered by the 51 EECs operating in the country (Greek Ministry of Education 2014). As an example, we present an educational programme offered by the EEC of Krestena to grades 7 and 8 of local middle schools. The programme is about the effects of river damming and was developed over the 2011–2013 school years when the first author of this chapter was a member of the centre's pedagogical team.

A single-day educational programme offered by the EEC of Krestena to secondary school students includes a visit to a local hydroelectricity dam and its environs, where students are invited to observe the apparent and implicit effects of damming on the natural environment. Through this process the students realize that a hydro-dam is an intervention with cross-scale, interconnected and ambivalent consequences. A hydro-dam disrupts the local ecosystem, but it also generates electricity and thus saves fossil fuels and restrains global carbon emissions. Dams can cause the displacement of local populations, loss of cultural heritage and soil erosion; yet, they serve as reservoirs of freshwater necessary for irrigation and urban use. At the closing of the educational programme, students are asked to debate and reach an informed decision on a hypothetical proposal for the construction of a similar dam in their area of origin. The hydro-dam programme draws on diverse disciplines (i.e. geography, geology, engineering, ecology) to expose the complexity of environmental issues and the role of conflicting interests in any major or minor environmental intervention.

Other EE programmes in Greece have ventured further to incorporate local knowledge about the evolution of the ecosystems and human communities, following the guidelines of UNESCO on cultural and ecological diversity (UNESCO–UNEP 2002). Recently, these efforts to integrate history, culture, mythology and ancient spirituality into a holistic approach of EE have attracted the interest of international organizations such as CEDEFOP (2012).

9.2 The Impact of the Economic Crisis on EE Infrastructure

As the EEC network is an organic and inextricable part of the modus operandi of EE in Greece, their phases of expansion, crisis and recovery reflect the comprehensive history of EE in the country (Fig. 9.2). During the lag phase, between 1993 and 2004, EECs functioned within selected administrative districts, and their dissemination was not enough to cover the EE needs of the educational community. However, this period was critical for the development of their institutional, operational and educational characteristics. Then, a phase of exponential growth fuelled by a favourable politico-economic environment saw the number of EECs increase fourfold within 5 years (2004–2009). During this period, EECs disseminated their work through peripheral, national and international thematic networks (Greek Ministry of Education 2005). A peripheral thematic network connects schools and institutions to address a peripheral issue, such as pollution in the Corinthian gulf. Respectively, an international thematic network works on an international environmental issue, such as overfishing in the Mediterranean.

Shortly after the global financial crisis of 2007–2008 and its' effect on the Greek economy, EE was targeted by policymakers as an ideal field for cutbacks. Promptly, the Greek government abolished the positions of EE counsellors, closed 40% of the EECs and curtailed the staff of the remainder (Greek Ministry of Education 2011). As a result, only half of the EE positions were retained after the cutbacks (Yanniris

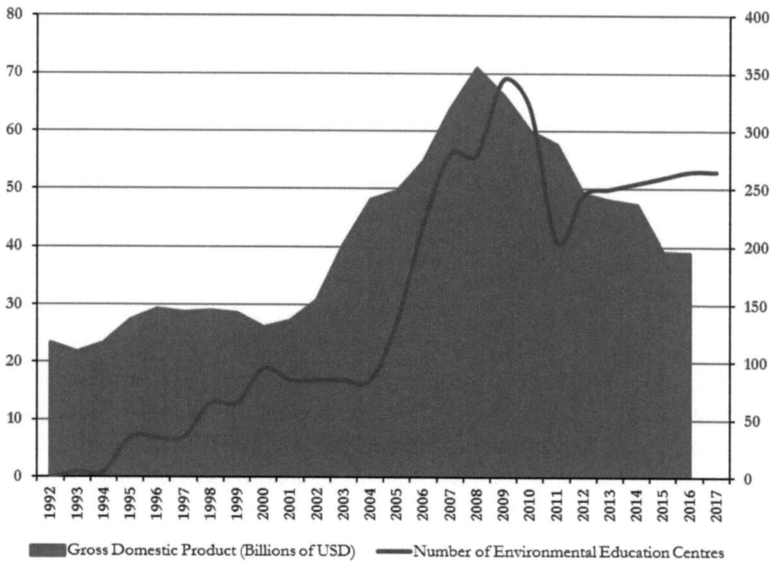

Fig. 9.2 Juxtaposition of the Greek GDP (World Bank 2017) with the number of Environmental Education Centres (EECs) in operation

2012). The political decision for the shrinking of EE waged criticism throughout the media, local governments and the national parliament (Notharos 2010). Overall, it was unclear whether the cutbacks have had a positive net effect on the public budget because the closure of EECs entailed an immediate loss of the respective EU funding. Following 2011, this trend began to reverse; the EE infrastructure started to show signs of recovery despite the continuing decline of the national economy (Fig. 9.2). Advocacy and political pressure resulted in the reopening of one-third of the EECs that were abolished in the wake of the crisis (Kouvelis et al. 2010); moreover, new EECs that were held in the pipeline started to appear (Greek Ministry of Education 2015). This positive outcome resulted from a broader mobilization in support of EE that involved teachers' associations, education officers, local communities and political parties. Local governments also lobbied against the closure of EECs, which they viewed as hubs of educational and economic activity for their localities (A. Xylokota, Greek Ministry of Education EE office director 2010–2015, personal communication, January 29, 2015).

9.3 Structural Characteristics of Resilient Systems

Ultimately, from a global perspective and within the purpose of this book, it is important to go beyond the local politics of the Greek case of crisis and recovery to identify what kind of multilevel governance and strategic choices have contributed to the resilience of this particular EE framework. If we are able to indicate that

Table 9.1 Systemic properties of environmental education infrastructure

Organizational scheme	Central	⟷	Peripheral
Content	Science based	⟷	Interdisciplinary
Practice	Indoor	⟷	Outdoor
Economic model	Private	⟷	Public
Access	Restricted	⟷	Open
Systematic assimilation	Embedded	⟷	Extracurricular
Participation	Required	⟷	Elective
Consistency	Continuing	⟷	Intermittent

certain structural characteristics add to the resilience of EE frameworks, other national contexts can capitalize of the Greek experience to avoid a similar loss in EE materials, know-how and infrastructure and build resilient EE systems. Resilience is described as the ability of complex systems to absorb external disturbance and to recover from disruptive change (Ehrlich 1986). According to David Orr (1992), resilience requires small, locally adaptable, resource-conserving and culturally suitable solutions. He proposed the following six criteria as prerequisites of resilience for any composite system: (i) modular, dispersed structure, (ii) multiple interconnections between components, (iii) short linkages, (iv) redundancy, (v) simplicity and (vi) loose coupling of components in a hierarchy (Orr 1992, p. 34).

The measure for resilience of any composite system is its ability to recover after its exposure to a major perturbation (Myers 1996). Respectively, the institutional EE framework that operates within a national context can be viewed as a composite system with its own structural properties that enhance or undermine its resilience. In the following section, we seek to parameterize the elemental properties of national EE frameworks through the systematic classification of eight bipolar qualities (Table 9.1). Then, we go over these parameterized systemic properties as applied to the Greek case and attempt a qualitative correlation with Orr's criteria. Based on the parameters of Table 9.1, we discuss the effects of the Greek EE systemic properties on its ability to recover from disruptive change. Eventually, this chapter aims to provide a more general insight to the strategic choices that enhance the resilience of EE providers.

9.4 Strategic Choices for Resilience in Environmental Education

Since its very beginnings in the early 1990s, EE in Greece has followed a decentralized organizational scheme. Provision was made for the horizontal appointment of EE counsellors throughout the country's 52 administrative districts (Greek Ministry of Education 1990). Furthermore, small, equally sized EECs were dispersed throughout the country (Fig. 9.1). Even though 40% of the EECs were closed after the crisis (Fig. 9.2), the remaining centres retained the resources and connections needed for the system to recover. Thus, the infrastructure of EE survived the crisis event and regenerated its damaged components. The geographical dispersal and redundancy of EE providers in Greece alleviated the risk of a generalized collapse. According to Orr (1992), a modular, dispersed structure is an essential requirement for resilience (criterion i). Moreover, Orr (1992) projected that the redundancy of components (criterion iv) is equally important for resilience as it diminishes the chance of an entire class of systemic components becoming extinct.

Furthermore, decentralization as a strategic choice entails a return to the local level (re-localization) and enables the use of locally available resources and local knowledge. Through re-localization and the shortening of food and energy supply lines, the system saves energy and resources (criterion iii). The strategic choice of dispersing EECs to the country's periphery enabled them to form shorter – and stronger – connections with local natural and human resources. Short linkages are less vulnerable and hence systemic resilience was enhanced (Orr 1992). Autonomy is an additional derivative of re-localization. The condition of autonomy is equivalent with a loose coupling of components in a hierarchy (criterion vi), an organizational scheme in which each of its components has, or makes use of, little or no knowledge of the definitions of other separate components (Weick 1976). In Greece, EECs enjoy an enhanced level of pedagogical autonomy, and each centre is free to develop its curricula, funding sources and affiliations (Tsaliki and Paradeisanos 2000).

Related with autonomy is the ongoing discussion on whether EE should be offered as part of the science curriculum or as a separate entity, entirely interdisciplinary in content (Brügger 2004). In the Greek system, an interdisciplinary approach to EE is enacted through legislation that regulates the implementation of EE programmes and procedures of EEC staffing. Specifically, each EEC is staffed by (five to six) teachers of various specialties from both primary and secondary education (Greek Ministry of Education 2011). The educational staff of EECs is encouraged to find thematic connections between disciplines and use these in the development of their educational programmes. Also, thematic networks between EECs, schools, non-governmental organizations and other environmental education providers create multiple interconnections between EE providers (Greek Ministry of Education 2005). According to Orr (1992), multiple interconnections between components is a prerequisite of resilience (criterion ii). In the event of crisis, it is the rearranging of these connections that compensates for any lost information and enables the system to retain its homeostasis.

As an educational philosophy, outdoor learning is essential to several of the currents of EE (Sauvé 2005). Accordingly, EE in Greece has a strong outdoor component that entails simplicity of pedagogical praxis. In the outdoors, teachers and students are disconnected from the transportation and energy grids and have to rely on simple and modest means and materials. Simplicity is a prerequisite for resilience according to the fifth (v) criterion set by Orr (1992).

During the last decade, there has been a growing understanding on the importance of cultural identity, local knowledge and place-specific skills in the field of EE (Yanniris 2016). Outdoor, place-based pedagogy connects local environmental issues with students' direct experience of their region and community. The knowledge of how to live (and how to live sustainably) in a particular place includes elements of local history, arts, culture oral tradition and, when applicable, indigenous heritage (Traina 1995, p. 2). Hence, outdoor practice is believed to allow students and instructors to make interdisciplinary links and to reconnect with their immediate natural environments.

A critical issue for EE is the funding sources for its activities. In Greece, EE is publicly funded through a multilateral scheme that involves local governments, the national government and the European Union (Greek Ministry of Education and EU 2010). A cross-level diversification of funding concurs with Orr's (1992) resiliency criteria i, ii and vi. Indeed, the multilevel sharing of funding has played a positive role in the resilience of EE frameworks in Greece. A strong argument used against the abolition of EECs has been that any savings at the level of the central government would be immediately cancelled by the loss of EU funds directed to the local economies through the EECs (Yanniris 2012).

Further diversification of funding sources could include a stronger presence for the private sector. Yet, there is no guarantee that a scheme based exclusively on private initiative would result in more resilient EE systems. For example, privatization of EECs in the UK has led to a decrease in the number of specifically trained and qualified staff (Higgins and Kirk 2006). A possible solution would be a scheme that retains public control over the guidelines and policies of EE while ceding operational services such as lodging and transportation to the private sector. Such a model could enhance the diversity of funding sources and protect the system from recurring crises. However, a certain level of participation by the public sector is desired in order to act as a safeguard of equitable access to EE (Yanniris 2015).

Uninhibited access to EE is important for vulnerable groups that experience physical, cultural, conceptual and financial barriers to their participation in outdoor educational programmes (Ham and Sewing 1988). Choices of inclusion or exclusion from EE ultimately depend on the policies and finances of EE providers. In Greece, lodging, alimentation and educational services are provided free of charge for the recipients of EE (Greek Ministry of Education and European Union 2010). EE is covered by a constitutional provision for free and open access to all levels of education (Greek Constitution art. 16, § 4). The extent of access to EE correlates with the effectiveness and resilience of EE systems. On a social level, open access amplifies the connections between the stakeholders of EE and relates with Orr's (1992) ii, iii and iv criteria for resilience. Hence, uninhibited access to EE, apart from being a right, an obligation and a choice of self-preservation, has a positive impact on the resilience of EE frameworks.

Two schools of thought have emerged with respect to the systematic assimilation of EE: those who see EE best provided as an extracurricular subject and those who advocate an embedded approach (Lynch et al. 2004). The embedment of EE as a standard subject of the curriculum would render it a more enduring status. Nevertheless, there is the counter argument that the incorporation of EE might jeopardize its innovative character and reduce it to just another indoor course deprived of the momentum or the sense of urgency that is needed to tackle the environmental problem (Russell et al. 2000). In Greece, EE continues to be taught as an independent, special topics subject (Yanniris 2015). Recently, the Pedagogical Institute of Greece (2011) has proposed a plan for the embedment of EE in the standard curriculum of secondary education. A qualitative analysis based on the table of systemic properties (Table 9.1) suggests that the choice of embedment would initially increase the resilience of EE as it would be directly connected with the permanent structure of the national education system. However, in the long-term EE could lose the resilience generated by autonomy of its components (criterion vi). Hence, with respect to the embedment of EE to the standard curriculum, there are two conflicting effects on the resilience of EE systems. Overall, it is questionable whether EE can retain the autonomy of its components as embedded in the standard curriculum. Much will depend on the strategy of embedment and how this will affect the other systemic properties of EE.

Despite all the institutional and financial support that EE has received, it has only involved a minority of students. A lack of teachers eager to include EE in their teaching has been reported. In Greece, pertinent research indicates that only one-third of in-service teachers included EE during instructional time (Mavrikaki et al. 2004). Teachers cite increased workload (52%) and their lack of training on EE (28%) as their major deterrent (Yanniris 2015). At the same time, 64% of the same sample of in-service teachers calls for EE to become compulsory (ibid). A shift of policy towards compulsory attendance to EE programmes for students would enhance participation and thus the resiliency of EE systems in accordance with Orr's (1992) criteria ii, iii and iv. However, this should be done in a manner that would protect the autonomy of EE, as described by criterion vi.

The last of systemic properties that we explore refers to the thematic, tropic and temporal continuity of EE. Presently, there are no safeguards in the Greek system for the consistency of EE between grades, practices or subject matters. Environmental projects are initially identified by teachers that introduce EE to their students, and then the students complete a specific project. These projects are completed within one school year, and their continuation depends on the availability of eager teachers. Hence, the present modus operandi of EE in Greece ingrains the risk that any class of students who followed EE in a single grade of primary school may either never encounter EE again in their schooling time or may re-engage in EE sporadically in higher grades or in secondary education. Orr (1992) has not included temporal or conceptual consistency as a criterion of systemic resilience, but it is understood that any educational process needs to be continuous in order to be effective. The fragmented delivery of EE in Greece suggests that certain steps need to be taken to ameliorate its coherence and secure the advances made in the field.

9.5 A Way Forward: Recommendations for the Greek Environmental Education System

The bibliography we consulted indicates that during the last three decades, Greece has acted as an experimental field for EE where different pedagogical techniques, practices and policies were tested. However, EE in Greece has spent excessive time in an experimental phase that restricted its practice to a finite group of committed teachers (Tsaliki 1998, pp. 188–189). Student participation in EE was affected by the reluctance of the country's political order to embed its content to the standard curriculum. Part of this lag can be attributed to the strict timetable and the centralized nature of the Greek educational system that have inhibited the integration of EE as an overarching and comprehensive activity (ibid).

Overall, in order to broaden the audience and enhance its resilience, EE needs to be ushered under the protective umbrella of formal schooling. The introduction of EE in the standard curriculum will allow it to become a required subject with universal participation. This also enables provision on the continuity of EE from grade to grade. After more than 25 years of systematic practice in Greece, EE is ripe to become an embedded, required and consistent course for primary and secondary schools. The distinct character of EE can be protected by observing the qualities of decentralization, interdisciplinarity and outdoor practice. Equitable access must be preserved, yet the transfer of a limited scope of EE services to the private sector can be considered as a way to strengthen its relationship with the local communities.

9.6 Conclusive Remarks

This chapter examined the structural components that contributed to the resilience of EE infrastructure in Greece. For all its weaknesses on the issues of participation, content assimilation and continuity, EE in Greece displays important assets, including an innovative network of small, independent, locally adaptable Environmental Education Centres. These characteristics generated the resilience of the EEC network through a severe economic crisis that caused the interruption and discontinuity of several structures in Greece. Overall, the practice of EE in the country meets the major theoretical requirements for resilience, which explains the heretofore recovery. The future behaviour of the Greek system will depend on its capacity to settle the following dilemmatic questions: (a) secure the funding of EE without jeopardizing equitable access for its recipients and (b) embed EE into the standard curriculum without ceding its innovative character. These dilemmas bear existential threats for EE and are very similar to the challenges that the field faces internationally. As the present chapter shows, EE can overcome difficulties if it adheres to its core principles. Resilience derives from the qualities of outdoor practice, decentralization and interdisciplinarity. Thus, for EE, the resilience of its infrastructure is a measure of consistency between theory and praxis.

Acknowledgement This research was supported by a Stavros S. Niarchos Foundation grant administered by the McGill Graduate and Postdoctoral Studies Office.

Questions

Discussion questions that could help instructors and students to engage in meaningful conversation about the ideas presented in this chapter.

1. According to the provided experiences, how is the EE infrastructure affected by the cycles of economic growth and recession? Does this pattern relate to your situation in any way?
2. Which strategic choices can enhance the resilience of EE infrastructure?
3. How can EE programmes become embedded into the standard curriculum without ceding their innovative character?
4. How can the funding of EE be secured without jeopardizing equitable access for its recipients?
5. Which parts of the article suggest political involvement in EE decision-making? Do you think it is possible (or desirable) to keep politics out of decision-making in education?

References

Brügger, P. (2004). 25 years past Tbilisi: Environmental teaching or cheating? In W. L. Filho & M. Littledyke (Eds.), *International perspectives in environmental education* (Vol. 16, pp. 129–138). Frankfurt: P. Lang.

CEDEFOP. (2012). Study Visit Group Report: *Environmental education centres: Supporting actions towards sustainability*. Retrieved from http://studyvisits.cedefop.europa.eu/assets/upload/documentation/SV_Group_reports/74_GR_GroupReport_Oct12.pdf

Ehrlich, P. R. (1986). *The machinery of nature*. New York: Simon & Schuster.

Flogaitis, E., & Alexopoulou, I. (1991). Environmental education in Greece. *European Journal of Education, 26*(4), 339–345.

Gardeli, S. (1983). The protection of our coasts. *Milieu, 15*, 6–7.

Gardeli, S. (1986). Education et environnement en Grèce: une expérience pilote "école et communauté". *Pour une Éducation à L'environnement, 2*, 56–65.

Georgopoulos, A. (2014). *Environmental education: Identity issues*. Athens: Gutemberg.

Greek Ministry of Education. (1990). Act 1892/1990 (ΦΕΚ 101 τ. Α), article 111 par. 13 (in Greek).

Greek Ministry of Education. (2005). *Thematic networks in environmental education.* Decree 66272/Γ7/04–07–2005/ΥΠΕΠΘ (In Greek).

Greek Ministry of Education. (2010). *Update of the programs provided by environmental education centers for the 2010–2011 school year.* Directive 157598/Γ 713–12–2010 (In Greek).

Greek Ministry of Education. (2011). *Procedures of appointment and responsibilities of teachers in environmental education centers.* Ministerial Decree 83691/Γ7/22–7–2011 (In Greek). Retrieved from https://www.minedu.gov.gr/publications/docs2011/diadikasies_kpe_ya_110725.pdf

Greek Ministry of Education. (2014). *Update of the programs provided by environmental education centers for the 2014–2015 school year.* Directive 208971/Δ2/22–12–2014 (in Greek). Retrieved from https://app.box.com/s/ix0va5iypsf8kkuaep7a

Greek Ministry of Education. (2015). *Establishment of the environmental education center of Delta, Thessaloniki.* Ministerial Decree 133936/Δ2/27–08–2015 (In Greek). Retrieved from http://dide.koz.sch.gr/tmsdr/wp–content/uploads/kpe–delta.pdf
Greek Ministry of Education and European Union. (2010). *Call for proposals for the operational plan 'Environmental education centers: Activities for students'.* Document 11941/23–8–2010 (In Greek).
Greek Ministry of Education and European Union. (2013). *Lifelong learning actions for sustainability education,* Act 19791/27–9–2013, ΑΔΑ: ΒΛ9Α9–ΟΝΡ (in Greek). Retrieved from http://www.edulll.gr/?p=16144
Ham, S. H., & Sewing, D. R. (1988). Barriers to environmental education. *The Journal of Environmental Education, 19*(2), 17–24.
Higgins, P., & Kirk, G. (2006). Sustainability education in Scotland: The impact of national and international initiatives on teacher education and outdoor education. *Journal of Geography in Higher Education, 30*(2), 313–326.
Katsakiori, M., Flogaitis, E., & Papadimitriou, V. (2008). *Present condition of environmental education in Greece–Environmental education centres.* Athens: EKEBY. (In Greek).
Kontaras, V. (2004). Identification of the basic characteristics of environmental education centers. Post–graduate thesis, Environmental Studies, University of the Aegean, Mytilene (In Greek).
Kouvelis, F., Leventis, A., & Tsoukalis, N., Psarianos, G. (2010). *Question to the minister of education in relation to the lack of funding for environmental education centers,* session B, period ΙΓ, Greek Parliament Proceedings.
Lynch, P. M., Leonard, R. L., & MacDonald, I. (2004). Identifying environmental education content in higher education: A New Zealand perspective. In W. L. Filho & M. Littledyke (Eds.), *International perspectives in environmental education* (pp. 138–152). Frankfurt and Main: Peter Lang.
Mavrikaki, E., Kyridis, A., Tsakiridou, E., & Golia, P. (2004). Greek educators' attitudes and beliefs about the application of environmental education in elementary school. In W. Leal Filho & M. Littledyke (Eds.), *International perspectives in environmental education* (pp. 29–36). Frankfurt and Main: Peter Lang.
Myers, N. (1996). Environmental services of biodiversity. *Proceedings of the National Academy of Sciences, 93*(7), 2764–2769.
Notharos, M. (2010). Padlock on green schools, *Eleftherotypia Newspaper,* (in Greek). Retrieved from http://www.enet.gr/?i=news.el.article&id=200007
Ntarzanou, K. (2008). *Environmental education in the second grade.* Unpublished undergraduate thesis, Harokopio University, Greece, 20–25 (in Greek).
Orr, D. W. (1992). *Ecological literacy: Education and the transition to a postmodern world.* New York: State University of New York Press.
Palmer, J. (1998). *Environmental education in the 21st century: Theory, practice, progress and promise.* New York: Routledge.
Pedagogical Institute of Greece. (2011). *Guide for the implementation of the 'Education for the Environment and Sustainability' program of studies in junior high school.* Retrieved from http://users.sch.gr/organopoulos/ekpaideytiko_yliko/odigos_eaa_gymn_2011.pdf.
Russell, C. L., Bell, A. C., & Fawcett, L. (2000). Navigating the waters of Canadian environmental education. In T. Goldstein & D. Selby (Eds.), *Weaving connections: Education for peace, environment and social justice* (pp. 196–217). Toronto: Sumach Press.
Sauvé, L. (2005). Currents in environmental education: Mapping a complex and evolving pedagogical field. *Canadian Journal of Environmental Education (CJEE), 10*(1), 11.
Traina, F. (1995). What is bioregionalism. F. Traina, S. Darley-Hill, Perspectives in bioregional education (1–12). Troy: North American Association for Environmental Education.
Tsaliki, E. (1998). The global scene: Greece. In J. Palmer (Ed.), *Environmental education in the 21st century: Theory, practice, progress and promise* (pp. 188–190). New York: Routledge.
Tsaliki, E., & Paradeisanos, A. (2000). The environmental education centres in Greece: The case of the Thessaloniki Centre in promoting environmental education in an urban industrial area. *Journal of Environmental Protection and Ecology, 1*(4), 429–435.

UNESCO–UNEP. (2002). Cultural diversity and biodiversity for sustainable development. World Summit on Sustainable development, Johannesburg, Retrieved from http://unesdoc.unesco.org/images/0013/001322/132262e.pdf

Weick, K. (1976). Educational organizations as loosely coupled systems. *Administrative Science Quarterly, 21*(1), 1–9.

World Bank. (2017). *Gross domestic product indicator*. Retrieved from http://data.worldbank.org

Yanniris, C. (2012). Is there a future for environmental education? *En Ethria Environmental Journal, 226*, 8–9. (In Greek).

Yanniris, C. (2015). 20+ years of environmental education centers in Greece: Teachers' perceptions and future challenges. *Applied Environmental Education and Communication, 14*(3), 149–166. https://doi.org/10.1080/1533015X.2015.1067578.

Yanniris, C. (2016). Place-based learning as a catalyst for change. *Pathways: Ontario Journal of Outdoor Education, 29*(1), 11–15.

Chapter 10
Storied Environmental Curriculum: A Case-Based Perspective on Environmental Education

Alandeom W. Oliveira

Abstract Narrative curricular practices constitute an often neglected feature of environmental education. In response to this issue, the present chapter examines the text structure (i.e., written organization) of two cases developed by biology teachers in Brazil. A narrative analysis revealed that (1) both cases shared common features such as a focus on relatively powerless victims (poor, illiterate, and young) in need of help and tragic plots involving literary representations of death and fear and (2) they differed with regard to features such as structural organization (open vs. closed narrative structures) and positioning of students (external bystanders vs. science experts in charge of solving a problem). The main significance of this study is that it highlights the need for educators to better understand what it means to develop a storied curriculum as well as the implications of narrativizing environmental issues in particular ways.

10.1 The Nonnarrative Environmental Curriculum

> Gone are the rich, metaphor–laden narratives that connected traditional societies to the place/spirit/ecology that provided them with sustenance… [Contemporary schools] reinforce the decontextualized, anthropocentric construct of Western individuals. (Sandlos 1998, p. 1)

In the above quotation, John Sandlos (1998) highlights the problematic state of formal environmental education as an endeavor that has become increasingly devoid of narrative activity. Unlike indigenous groups in various parts of the world who have traditionally resorted to storytelling as a powerful tool for enculturating youth into particular ways of making sense of and interacting with nature (Basso 1996), formal environmental instruction is characterized by reduced narrative engagement

A.W. Oliveira (✉)
Educational Theory and Practice Department, State University of New York at Albany,
1400 Washington Ave., ED 113B, Albany, NY 12222, USA
e-mail: aoliveira@albany.edu

G. Reis, J. Scott (eds.), *International Perspectives on the Theory and Practice of Environmental Education: A Reader*, Environmental Discourses in Science Education 3, https://doi.org/10.1007/978-3-319-67732-3_10

and activity. When learning about the environment, students engage predominantly with the traditional science textbook, an impersonal written genre that favors factual exposition and context-independent generalization (Maton 2013). Exclusive exposure to this nonnarrative environmental curriculum, Sandlos (1998) argues, can encourage students to see themselves as placeless beings detached from the environment and develop amoral relationships with the natural world.

Nonetheless, narrative practice is not completely absent from science classrooms. In recent years, a growing number of educators have resorted to environmental cases – written texts that provide a storied account of problematic environmental situations. Unlike traditional textbooks, cases are short texts with spatiotemporally specific and engaging narrative stories with characters, dialogue, and a plot (Derriso 2011) designed to promote student learning by means of realistic, engaging, and contextualized written texts. Rather than serving as repository of orally based ways of knowing (traditional knowledge) or transmission of cultural values, environmental cases are usually designed for the pedagogical purpose of narrating environmental problems. That is, they introduce a nature-related discrepancy meant to engage a group of learners (problem solvers) in a search for a solution by fostering in them a need to eliminate the presented discrepancies (Arlin 1989). As problem solvers, students are driven by the goal of arriving at an acceptable or preferred solution, which typically requires a sophisticated problem schema, domain knowledge, general problem-solving skills, and reflective judgment. As David Jonassen (1997) writes, "problems are traditionally defined by a problem domain, a problem type, a problem–solving process, and a solution" (p. 66).

Unfortunately, as a direct consequence of the dearth of scholarly work in this area, environmental cases used by teachers tend to be laden with problematic understandings of nature that are anthropocentric and positivist, hence falling short of telling the "real story" of the centrality of nature to the sustainability of life on the planet. More than merely representing problematic situations through textually encoded factual information, environmental narratives invariably communicate social values and cultural meanings about nature shared by the storyteller's community. As Sandlos (1998) writes, "the environmental educator must never forget that the 'end' of the stories he or she tells is not merely the solution to a series of environmental problems, but the expression of a culture that has mastered the art of living well" (p. 7).

Given the above state of affairs, the present chapter examines two curricular narratives written by a group of biology teachers in Brazil. In doing so, an initial step is taken toward a more sophisticated, theory-based understanding of environmental education as storied curriculum development. Its appeal to an international readership stems from its potential to inform cultural curricular emplotment – how to effectively plot narrative curricula (cases) for specific cultural contexts.

10.2 How Teachers Use Cases

Environmental cases have been increasingly used to support student acquisition of scientific argumentation (Jiménez-Aleixandre and Erduran 2008). Designed to serve as springboards for evidence-based argumentation, cases narrativize environmental situations in particular ways. Troy Sadler et al. (2007) use two fictitious scenarios about (1) a coastal wildlife preserve near a shipping port and Native American tribes where fish and wild animal populations have considerably declined and (2) a small town faced with the decision on whether to replace a coal-burning power plant with a nuclear one. Alandeom Oliveira et al. (2012) resort to dilemmas such as a hypothetical encounter with a fawn in the woods, observation of classroom lights being unnecessarily left on, and the handling of an unwanted pet iguana. Katherine Kortenkamp and Colleen Moore (2009) use two environmental scenarios of ecologically damaging events: (1) a park manager whose decision to allow bike riders and dogs into a previously restricted park produces damage to wildlife and plants and (2) a pet owner who allows destructive cats to go outdoors and harm nearby birds. António Almeida et al. (2013) present children with scenarios involving animals: (1) the real killing of codfish-eating seals in Canada and (2) a family's hypothetical encounter with a hedgehog and a fox in a forest. These studies show that the presence of animal characters in the stories have a profound impact on students' environmental reasoning, usually prompting biocentric arguments – pro-environment stances based on appeals to the rights and intrinsic values of nature (as opposed to utilitarian arguments centered on human needs and interests).

Educators have also turned to case-based teaching methods in an effort to contextualize their teaching of abstract content. Clyde Herreid (2005) has students create and share their own cases on the topic of global warming based on their reading of Michael Crichton's novel "State of Fear." Jennifer Eastwood et al. (2012) employ socioscientific issues as contexts for nature of science instruction in high school biology. Cynthia Passmore and Jim Stewart (2002) resort to cases as a means to provide biology students with authentic and realistic contexts for exploring natural selection and constructing evolutionary explanatory models. As such, cases can serve as an effective alternative to decontextualized lecture-based transmission of generalized principles and concepts.

Lastly, cases have been used to teach controversial topics as varied as biotechnology (Dori et al. 2003), microbial cleanup of oil spills (Jones 2011), and birth control methods (Zojonc and Lavoie 2011). By strategically leveraging the controversy (ethical and moral) and real-life relevance of cases, these educators increased student motivation and interest.

10.3 Revealing the Structure of Stories

Stories have been shown to have an internal structure. Previous studies of stories in face-to-face conversation (Labov and Waletzkty 1967) revealed that oral reports of previous events or experiences share a common internal structure. The narrator usually begins by summarizing what the story is about (Abstract) and then identifying who was involved (e.g., characters), where the story took place, and when it occurred (Orientation). These initial phases are followed by chronological description of the event(s) that took place (Complicating Action), assessment of the significance or point of the story (Evaluation), identification of ultimate results or outcomes (Resolution), and closing off of the narration (Coda). Although these structural elements are sequentially and orderly placed in fully formed narratives, they can sometimes be implicit, embedded, and even missing from more complex narrative structures. For instance, although typically found at the end, Evaluation is sometimes dispersed throughout the narrative.

News stories from the press have been shown to be structurally similar to personal narratives, with a few noticeable differences. Allan Bell (1991) describes how in newspaper stories Evaluation is explicitly placed in the Abstract (rather than being concentrated at the end of the narrative), Complicating Action is reported in a non-chronologic fashion (cyclic temporal structure and sequencing of events based on perceived news value), and there is no Resolution (clear-cut outcome) or Coda (explicit signal that the storytelling has come to an end). Further, unlike personal narratives, printed stories have additional structural components such as Attribution (explicit identification of the story's source or origin) as well as Follow-Up, Commentary, and Background information; these are usually integrated with the events being described in the Complicating Action.

In scholarly fields such as folklore and narratology, evidence abounds that cultural influences can have profound effects on the production of narrative texts (Nash 1990). This is particularly evident in the occurrence of cultural masterplots (Abbott 2008), stories recurrently told and deeply embedded in a cultural context. Descriptive rather than prescriptive, this literature shows that despite their structural similarities, contextual factors such as mode of delivery, cultural influences, and emphasis on creativity can result in noticeable differences in storytelling format and organization.

10.4 Cases as Cultural Artifacts

This chapter is methodologically aligned with a tradition of anthropological research known as native ethnography (Russell 2002). Such an alignment stems primarily from the fact that the examined data were naturalistically collected in the same sociocultural context where the researcher grew up and was educated.

10.4.1 Teachers as Case Writers

A group of six in-service teachers (one male and five females) volunteered to participate in this study. These participants were recruited from a group of 30 high school instructors who taught biology in public high schools in a rural part of the state of Goiás (the Brazilian Midwest). These teachers were offered a 5-week professional development program on the use of case-based teaching at Goiás State University. Familiarity with case-based pedagogy served to ensure that participating teachers had sufficient practical expertise to plan their narrative curricula. Further, an effort was made to select experienced instructors (at least 10 years of teaching experience).

The participants wrote two environmental narratives. The first case "A Child in Distress" addressed the importance and risks of taking the measles–rubella vaccine (known in Brazil as "the double viral") during pregnancy. Entitled "What Is Wrong with Little José?," the second case tells the story of a poor boy who gets infected with schistosomiasis after swimming in contaminated water. This disease, commonly referred to as bloating or snail fever (a reference to its transmission vector), is caused by parasitic worms of the genus *Schistosoma* and typically affects human populations in developing areas that lack basic sanitation and infrastructure (for these cases, see below).

10.4.2 Brazilian Storytelling

Since text production is influenced by sociocultural context, I now provide background information about the cultural milieu in which teachers wrote their socioscientific cases. In his extensive sociological study of the local culture in Goiás, Oliveira (2006) describes the prevalence of social representations of fear and catastrophe in the local literary production, arts, and the media. Based on real events such as the Radioactive Accident of 1987 (Watts et al. 1997), these social representations convey an ambivalent sense of fear and admiration for technology and modernity in general, typically taking the form of "narratives whose function is to describe, analyze, or explain misfortune" (p. 297). Further, narratives of catastrophe tend to highlight the pain and suffering undergone by those victimized (the children, the poor, and the illiterate) as well as the actions of altruistic heroes and selfish villains in an exaggerated style reminiscent of Greek tragedy. Another important element of this cultural masterplot is social denouncing, a shared concern for critically commenting on social inequity and assessing responsibility and blame. As Oliveira (2006) points out, "moments of pain and suffering constitute a unique opportunity for taking ethical positions" (p. 299). Discussion of situations where individual or collective (in)action results in death (e.g., the 9/11 attack in the USA, Chernobyl disaster in Ukraine, etc.) conjures up a spectrum of responsibility-related emotions, such as blame, anger, and indignation about insufficient prevention, and hence sparks lively moral deliberation. This is precisely what the writers from this sociocultural literary tradition tend to do when composing.

Like in other parts of Brazil, traditional literary production in Goiás is characterized by a relatively strong emphasis on orality-based genres (Oliveira 2006) – texts written in a spoken style with pervasive use of conversational devices such as dialogues, colloquialism, etc. This is particularly evident in the high popularity of traditional texts meant to be delivered orally such as Cordel or string books (ABLC 2009) – booklets with poetic narratives and woodcut illustrations that are usually recited or sung publically and are sold to the general public by street vendors. Such a literary preference for popular orality has been recently explored by local science and health educators who set out to educate the public about diseases and science (de Oliveira et al. 2011). As described below, these cultural elements strongly influenced teacher text production.

10.4.3 Uncovering the Cases' Narrative Structures

To examine the structure of the two environmental cases, a narrative analysis was conducted. Both qualitative and comparative, this analytical approach is consistent with an interpretivist ethnographic study of texts (Russell 2002) whose goal is to uncover and better understand meaning in textual artifacts. Narrative analysis focuses specifically on the characterization of the written structure of teachers' cases through identification of its salient organizational features and narrative elements.

Because the cases invariably took the form of stories and showed little structural similarity with standard scientific articles, this analysis became guided by Labov's (1972) theoretical framework for the internal structure of narratives. More specifically, my analysis was focused on assessing how closely the investigative cases followed the narrative sequence of structural elements: Abstract – Orientation – Complicating Action – Evaluation – Resolution – Coda. Each structural element was then further analyzed through a finer comparative examination of internal narrative elements such as the specific types of characters, activity, setting, and plot used in the story.

10.5 The Storied Environmental Curriculum

Overall, "A Child in Distress" was longer and more structurally complex than "What Is Wrong with Little José?" (343 and 169 words, respectively). With a considerably more elaborate narrative structure, "A Child in Distress" involved a larger number of characters (an unborn child, mother, father, health agent, supervisor, doctor) and multiple scenes (at home, at the health agency, and at the hospital). On the other hand, "What Is Wrong with Little José?" had only two characters (José and his mother) and comprised a single scene (a visit to José's home).

Despite their varied length and focus, the two environmental cases shared a common three-part structure: Orientation→ Complicating Action→ Resolution. As described below, by not including macrostructural elements such as Abstract and Evaluation, teachers afforded students some degree of freedom (interpretative room) to define what the point of each story was as well as its significance.

Case 1: A Child in Distress

It was late at night when the phone rang. Startled, Cristina Torres answered the phone; she was receiving an emergency call from Mrs. Josefa Freitas. She was pregnant and asked for help: "Please Cristina, I'm sorry about the late hour, but I need you to come visit me right away, after all I am pregnant. My husband does not know yet and I am afraid of telling him everything that happened, because I have having a hemorrhage and it doesn't stop." Mrs. Josefa was the fifth pregnant woman on Cristina's files, but she was the first one to request an unscheduled visit. Cristina Torres was a health agent, while she finished her degree in biology.

Cristina gathered some papers, grabbed her materials, and went straight to Mrs. Josefa's house. Upon her arrival, she was surprised that the husband had just found out the truth and was very anxious and worried. He wanted by all means to take his wife to the hospital, but she resisted. After a while, the health agent convinced Mrs. Josefa to go to the hospital. Next day, Cristina approached their supervisor Maria de Jesus to discuss the situation. Indeed, both went to the hospital, and, to everyone's disappointment, they received the news that the mother miscarried the fetus.

During a talk with Dr. Alexandre, the city's gynecologist who examined Mrs. Josefa, they found out that the reason for her hemorrhage was a vaccine that Mrs. Josefa had recently taken. The vaccine was called MR which protects against measles and rubella. "I see now," said the supervisor, "there is a vaccination campaign ongoing this week and there were three cases yesterday, but the mothers did not lose their babies." The agent agreed with her supervisor regarding the populace's lack of information, and for this reason her supervisor told Cristina to have a meeting with the other agents to create a clear informative pamphlet about the topic. This pamphlet will be utilized to make the population more aware during the week preceding other vaccination campaigns.

Both narratives began with spatiotemporal specifications ("in a poor house next to yours") and introduction to important dramatis personae (Prop 1968), that is, characters by mention of names, description of attributes, and indication of status. In "What Is Wrong with Little José?," the initial situation involved an observer (a mother, the reader himself/herself) noticing something unusual about the health of a victim (a child suspected of being unwell or sick). By contrast, "A Child in Distress" begins by introducing readers to a pregnant woman who reaches out to a helper (a scientifically literate health agent). In both cases, teachers begin to set the scene for situations involving biologically deviant cases (Polanyi 1985), that is, stories about departures from the biological norm (e.g., unusual health conditions).

Although both narratives rely on fictional characters and made-up events inspired by science-related events commonly seen in the news (dramatization of science in the news), several differences are noticeable in the structural organization of the initial situations. First, while "What Is Wrong with Little José?" was written as a series of events unfolding in the present tense, "A Child in Distress" was written in the past tense. The former is focused on what is happening right now, whereas the latter narrates what has already happened. A second difference is that in "What Is Wrong with Little José?," students are addressed directly (in *your* town, in a poor house next to *yours*... one day *you* come over to visit). By contrast, "A Child in Distress" is a third-person narrative from which both writers and readers are completely excluded (i.e., stories about people other than the teacher and students). And third, "What Is Wrong with Little José?" introduces a layperson with a lower socioeconomic status (a poor child), whereas "A Child in Distress" presents a professional knowledgeable about science (a health agent).

After introducing the main characters and presenting an initial situation, each narrative entered a phase of complicating action. One type of complication present in both cases was a failed attempt to solve the problem at hand. Little José's mother attempts to cure him with homemade remedies, but his health continues to deteriorate. Likewise, in "A Child in Distress," the health agent fails to prevent the miscarriage and save the fetus' life despite her best efforts. This type of complicating action (failed attempts) creates the impression of an escalation of the original situation or problem. Not only do persistence of symptoms and worsening of illness suggest that the initial problematic conditions to a certain extent get worse over time, but they also convey a sense of increasing urgency. Action-taking and decision-making must occur quickly in order to arrive at a solution and potentially save a life.

Another noticeable theme in the two cases was death. Whether focused on premature or potential death, teachers invariably emphasized the interconnectedness of science and death in modern society. In close alignment with local cultural masterplots of fear and catastrophe (Oliveira 2006), one group of teachers wrote about dramatic instances involving the accidental death Mrs. Freitas' fetus in "A Child in Distress." In doing so, these teachers provided their student–readers with a situation wherein a layperson's illiteracy accidentally led to the premature death of an unborn. Premature and preventable endings of human life such as this, sociologists point out, have become associated with "bad death" (i.e., morally wrong demises) in modern societies with the advent of science and technological innovation. As Kearl (1989) writes "death has been the instrument routinely used to demarcate the moral boundaries of [modern] life" (p. 299). Similarly, by including the passing of a fictitious fetus in the narrative, teachers capitalized on premature death's potential to prompt classroom discussion of one's moral commitments with regard to the value of life and interrelationships between death and life as students negotiate the moral and ethical ramifications of death caused by science.

Case 2: What Is Wrong with Little José?
Little José is 11 years old and lives in your town, in a poor house next to yours. He likes to play and to sell Popsicle on the streets and he is not into studying much. One day you come over to visit Little José and notice that he does not seem well.

He continuously scratches all of his body, especially his legs. His mom gives him some homemade remedies that she prepared and the itching soon stops. Sometime later (about 1 month), you return to your little friend's house and find him discouraged and pale. His mother was saying to him: "You must exercise more, maybe play or walk. I even let you swim in the lake, as you have secretly done before. Otherwise you will get even fatter." It was then that you perceived that, besides his discouraged appearance and yellowish color, Little José's belly was swelled up as a balloon. Then, a question comes in your head: what is wrong with Little José?

In "What Is Wrong with Little José?," the potential death of humans due to infection by the pathogenic agent of schistosomiasis was indirectly connected to the polluting acts of other people. As a result, emphasis is placed on how man's killing of nature can ultimately result in the death of fellow human beings (i.e. the indirect death of man by man). By illuminating the specifics of such indirect causality, science now emerges as a valuable instrument for the prevention and mitigation of cases of environmentally mediated accidental death (Kearl 1989), that is, manmade human endings that lack intentionality.

In "A Child in Distress," the complicating action was followed by resolution, with the identification of a specific cause of the narrated problem, namely, a harmful substance (the MR vaccine). Resolution took the form of an official prognosis made by a science expert (a gynecologist) who held an authoritative social status. By contrast, "What Is Wrong with Little José?" remained unresolved. Like newspaper stories (Bell 1991), this open case concludes without closure or specific answers. Instead of authoritatively pronouncing a scientific prognosis, the pretext ends with a question addressed directly at the readers: "Then a question comes in *your* head: what it is wrong with Little José?" This question encourages the readers to take on the role of science experts who are entitled to make decisions and take action based on the available evidence and knowledge of scientific concepts (as opposed to relying on others' authoritative decision-making and action-taking).

Another important structural feature of teachers' investigative cases was their lack of an Evaluation. Unlike conversational storytellers who typically communicate a point or moral (Labov and Waletzky 1967), teachers invariably wrote narrative texts without such evaluative structural element. The lack of an Evaluation is very important given the pedagogical function of such written science stories. By neutrally narrating science-related events without expressing a particular evaluative stance, teachers encouraged students to take on the roles of evaluators whose task was to assess in writing the scientific and social significance of unfolding action in scenarios that were familiar and directly relevant to students' lives.

10.6 Curricular Materials as Social Acts

The above analysis revealed that both "A Child in Distress" and "What Is Wrong with Little José?" shared common features such as a focus on relatively powerless victims (poor, illiterate, and young) in need of help and tragic plots involving literary representations of death and fear but varied with regard to features such as structural organization (open vs. closed narrative structures) and positioning of students (external bystanders vs. science experts in charge of solving a problem). By reporting on these particular states of affairs, teachers defined particular situations and created social facts, understandings, and arrangements to be treated as true during subsequent pedagogical activity.

Such findings invite us to further theorize about the nature of case-based environmental education. Based on Charles Bazerman's (2004) argument that "texts do things" (i.e., have a social force), environmental narratives can be viewed as social acts whose performance enables teachers to accomplish particular tasks, organize their social activity, and take on particular social roles. One good example of this social force of texts is teacher lesson plans' power to shape science learning opportunities and influence student experiences. For instance, lesson plans written as contingency plans (flexible sequence of instructional phases) afford students more openness and self-direction than lesson plans written as detailed scripts (Abell 2004). From this perspective, the environmental cases written by teachers not only mean something (carry propositional content) but also do something (constitute particular types of social acts within an activity structure). Like speakers performing oral moves, teachers can be viewed as producing texts that have locutionary content (i.e., referential information) as well as illocutionary force (i.e., potential to accomplish specific types of pedagogical work). When a teacher composes an environmental narrative, she/he performs a signifying act that serves to (re)constitute culturally valuable social structures and systems of human activity (Bazerman 2004). As such, environmental narrative-based curriculum can be said to perform social work in the sense of being constitutive of structured pedagogical activity in science classrooms. The storied environmental curriculum constitutes a social act or effort that teachers performed with the intent of creating appropriate pedagogical conditions and shaping social infrastructure of classrooms in a manner that was conducive to the emergence of particular systems of human activity (e.g., meaning making).

Therefore, it can be argued that "What Is Wrong with Little José?" has the illocutionary force of a socially closed, convergent scientific problem with a single acceptable resolution, namely, provision of the right diagnosis. This is particularly evident in the fact that the title of the story itself takes the form of a display question (Oliveira 2010), a traditional "testing query" designed to afford students the opportunity to publically display their knowledge of the one and only right answer (i.e., schistosomiasis).

Such a prompt is likely to invite students to position themselves as science novices displaying their knowledge of facts accepted by the expert community (i.e., providing the right answer). This case offers readers a point of reference (an increasingly specific meaning orientation) rather than horizons of possibilities (i.e., room for open-ended and divergent exploration of meanings), two distinct knowledge orientations commonly found in text-centered classroom activity (Langer 2011).

By contrast, "A Child in Distress" has the illocutionary force of a larger social problem in the sense that it presents a problematic situation that, though scientifically closed, is still open to social resolution. Students are invited to address a common social issue, namely, scientific illiteracy and lack of awareness among the populace (as opposed to simply displaying their knowledge of their right answer). A variety of socially acceptable solutions can be pursued by the students. This openness is likely to afford students a degree of freedom to explore their own thinking with regard to socioscientific issues (Sadler 2004) as well as engage in personal action (Pedretti and Hodson 1995), that is, to do something about the issues being investigated (as opposed to simply talking about them). As such, this environmental narrative is more closely aligned with what Roland Barthes (1974) describes as writerly texts – nonrestrictive stories characterized by a plurality of meaning that afford a variety of possible interpretations and encourage readers to actively create meaning. As a result, subsequent systems of pedagogical activity seem more likely to surpass the scientific sphere and reach the personal and social spheres. Students have room to strategically draw upon cultural knowledge, values, beliefs, and attitudes and to take a moral and personal stand.

Another important structural feature of the two cases was the lack of an Evaluation. As pointed out above, unlike conversational storytellers who typically communicate a point or moral (Labov and Waletzky 1967), teachers invariably wrote narrative texts without such an evaluative structural element. This neutral writing style has the illocutionary force of encouraging students to take on the roles of evaluators whose task was to assess in writing the scientific and social significance of unfolding action in scenarios that were familiar and directly relevant to students' lives.

10.7 Toward a Transformative Narrative

> Narratives transform life's journeys to sequences of events and evoke shifting and enduring perspectives on experience. (Ochs and Capps 1996, p. 20)

In my attempt to better understand EE from a case-based perspective, I have emphasized the need for examining the narrative curricular practices of classroom instructors in more depth. In particular, the reported findings highlight what it means to develop a more transformative storied curriculum as well as the implications of narrativizing environmental issues in particular ways. Rather than simply providing students with a problem, environmental educators can foster transformation by resorting to cultural masterplots and strategically drawing upon social representations (Moscovici 2003), that is, culturally shared knowledge used by members of a society to construct meaning and understand reality in certain ways. Doing so can enable teachers elsewhere to disrupt dominant anti-narratives or anti-stories of economic progress and modernity (White 1987). Central to our current "story of ecological crisis," such a nonnarrative attitude reduces the natural world into a mere commodity and narrative form to one single and privileged interpretation based on dualistic separations of fact and value and human and nature. It is my hope that the reported findings can help educators recognize the transformative potential of narrativized curriculum and tell students a different story, one of plurality of meanings, agency, and connection with nature.

Questions

Discussion questions that could help instructors and students to engage in meaningful conversation about the ideas presented in this chapter.

1. What was your pedagogical assessment of the two environmental cases produced by the Brazilian teachers? What do you consider to be the strengths and weaknesses of their narrativizing efforts?
2. Do you agree with the main argument of this chapter that environmental narratives can indeed enrich and empower the school curriculum? In what ways?
3. How can environmental educators effectively capitalize on the pedagogical potential of traditional stories about nature?
4. Describe an environmental story that you have previously been told (perhaps as a child) and that you remember particularly well. What specific attributes of this story made it memorable to you (e.g. plot, characters, delivery, length, genre, etc.)?
5. Using resources such as local newspapers, identify an environmental issue relevant to your personal context. How would you go about narrativizing it for classroom use? What would your environmental case look like?

References

Abbott, H. P. (2008). *The Cambridge introduction to narrative* (2nd ed.). Cambridge: Cambridge University Press. https://doi.org/10.1017/cbo9780511816932.

Abell, S. K. (2004). Be prepared! In T. R. Koballa & D. J. Tippins (Eds.), *Cases in middle and secondary science education: The promise and dilemmas* (2nd ed., pp. 83–87). Upper Saddle River: Pearson Merrill Prentice Hall.

Academia Brasileira de Literature de Cordel (ABLC). (2009). *Ciências em versos de cordel.* Retrieved from http://www.ablc.com.br/projetos/ciencia/ciencia.htm

Almeida, A., Vasconcelos, C. M., Strecht-Ribeiro, O., & Torres, J. (2013). Non–anthropocentric reasoning in children: Its incidence when they are confronted with ecological dilemmas. *International Journal of Science Education, 35*(2), 312–334. https://doi.org/10.1080/09500693.2011.608387.

Arlin, P. K. (1989). The problem of the problem. In J. D. Simmott (Ed.), *Everyday problem solving: Theory and applications* (pp. 229–237). New York: Praeger.

Barthes, R. (1974). *S/Z: An essay*. New York: Hill & Wang.
Basso, K. H. (1996). *Wisdom sits in places: Landscape and language among the Western Apache*. Albuquerque: University of New Mexico Press.
Bazerman, C. (2004). Speech acts, genres, and activity systems: How texts organize activity and people. In C. Bazerman & P. Prior (Eds.), *What writing does and how it does it: An introduction to analyzing texts and textual practices* (pp. 309–339). Mahwah: Lawrence Erlbaum. https://doi.org/10.4324/9781410609526.
Bell, A. (1991). *The language of news media*. Oxford: Blackwell.
de Oliveira, P. M. P., Paiva, J. S. P., Cezario, K. G., & Pagliuca, L. M. F. (2011). Literatura de cordel como estratégia educativa para prevenção da dengue. *Texto & Contexto Enfermagem, 20*(4), 766–773. https://doi.org/10.1590/S0104-07072011000400016.
Derriso, A. (2011). Teaching forward: Using case–based instruction to foster critical thinking. *The Science Teacher, 78*(2), 48–49.
Dori, Y. J., Tal, R. T., & Tsaushu, M. (2003). Teaching biotechnology through case studies – Can we improve higher order thinking skills of nonscience majors? *Science Education, 87*(6), 767–793. https://doi.org/10.1002/sce.10081.
Eastwood, J. L., Sadler, T. D., Zeidler, D. L., Lewis, A., Amiri, L., & Applebaum, S. (2012). Contextualizing nature of science instruction in socioscientific issues. *International Journal of Science Education, 34*(15), 2289–2315. https://doi.org/10.1080/09500693.2012.667582.
Herreid, C. F. (2005). Using novels as bases for case studies: Michael Crichton's state of fear and global warming. *Journal of College Science Teaching, 34*(7), 10–11.
Jiménez-Aleixandre, M. P., & Erduran, S. (2008). Argumentation in science education: An overview. In S. Erduran & M. P. Jiménez-Aleixandre (Eds.), *Argumentation in science education: Perspectives from classroom–based research* (pp. 3–28). New York: Springer. https://doi.org/10.1007/978-1-4020-6670-2.
Jonassen, D. H. (1997). Instructional design models of well-structured and ill-structured problem-solving learning outcomes. *Educational Technology Research and Development, 45*(1), 65–94. https://doi.org/10.1007/BF02299613.
Jones, D. (2011). Case study: Using microbe molecular biology for gulf oil spill clean up. *Biochemistry and Molecular Biology Education, 39*(2), 157–164. https://doi.org/10.1002/bmb.20471.
Kearl, M. C. (1989). *Endings: The sociology of death and dying*. New York: Oxford University Press. https://doi.org/10.1017/s0144686x00004487.
Kortenkamp, K. V., & Moore, C. F. (2009). Children's moral evaluations of ecological damage: The effect of biocentric and anthropocentric intentions. *Journal of Applied Social Psychology, 39*(8), 1785–1806. https://doi.org/10.1111/j.1559-1816.2009.00504.x.
Labov, W. (1972). *Language in the inner city: Studies in the black English vernacular*. Philadelphia: University of Pennsylvania Press.
Labov, W., & Waletzky, J. (1967). Narrative analysis. In J. Helm (Ed.), *Essays on the verbal and visual arts* (pp. 12–44). Seattle: University of Washington Press.
Langer, J. A. (2011). *Envisioning knowledge: Building literacy in the academic disciplines*. New York: Teachers College Press. https://doi.org/10.5860/choice.49-1010.
Maton. (2013). Making semantic waves: A key to cumulative knowledge–building. *Linguistics and Education, 24*(1), 8–22. https://doi.org/10.1016/j.linged.2012.11.005.
Moscovici, S. (2003). *Representações sociais: Investigações em psicologia social*. Petrópolis: Vozes.
Nash, C. (Ed.). (1990). *Narrative in culture*. London: Routledge.
Ochs, E., & Capps, L. (1996). Narrating the self. *Annual Review of Anthropology, 25*(1), 19–43. https://doi.org/10.1146/annurev.anthro.25.1.19.
Oliveira, E. C. (2006). *As representações do medo e das catástrofes em Goiás*. Unpublished doctoral dissertation, Universidade de Brasília. oai:repositorio.bce.unb.br:10482/1976
Oliveira, A. W. (2010). Improving teacher questioning in science inquiry discussions through professional development. *Journal of Research in Science Teaching, 47*(4), 422–453. https://doi.org/10.1002/tea.20345.

Oliveira, A. W., Akerson, V. L., & Oldfield, M. (2012). Environmental argumentation as sociocultural activity. *Journal of Research in Science Teaching, 49*(7), 869–897. https://doi.org/10.1002/tea.21020.
Passmore, C., & Stewart, J. (2002). A modeling approach to teaching evolutionary biology in high schools. *Journal of Research in Science Teaching, 39*(3), 185–204. https://doi.org/10.1002/tea.10020.
Pedretti, E., & Hodson, D. (1995). From rhetoric to action: Implementing STS education through action research. *Journal of Research in Science Teaching, 32*(5), 463–485. https://doi.org/10.1002/tea.3660320505.
Polanyi, L. (1985). *Telling the American story: A structural and cultural analysis of conversational storytelling*. Norwood: Ablex. https://doi.org/10.1525/aa.1986.88.4.0 2a00830.
Prop, V. (1968). *Morphology of the folktake*. Austin: University of Texas Press.
Russell, B. H. (2002). *Research methods in anthropology: Qualitative and quantitative approaches* (5th ed.pp. 443–449). Walnut Creek: AltaMira Press.
Sadler, T. D. (2004). Moral and ethical dimensions of socioscientific decision–making as integral components of scientific literacy. *Science Educator, 13*(1), 39–48.
Sadler, T. D., Barab, S. A., & Scott, B. (2007). What do students gain by engaging in socioscientific inquiry? *Research in Science Education, 37*(4), 371–391. https://doi.org/10.1007/s11165-006-9030-9.
Sandlos, J. (1998). The storied curriculum: Oral narratives, ethics, and environmental education. *The Journal of Environmental Education, 30*(1), 5–9. https://doi.org/10.1080/00958969809601857.
Watts, M., Alsop, S., Zylbersztajn, A., & Silva, S. M. (1997). 'Event–centred learning': An approach to teaching science technology and societal issues in two countries. *International Journal of Science Education, 19*(3), 341–351. https://doi.org/10.1080/0950069970190306.
White, H. (1987). *The content of the form: Narrative discourse and historical representation*. Baltimore: Johns Hopkins University Press. https://doi.org/10.2307/1773072.
Zojonc, S., & Lavoie, B. (2011). Addressing misconceptions about birth control: Case studies immersing students in the facts and real–life decisions. *The American Biology Teacher, 73*(6), 353–356. https://doi.org/10.1525/abt.2011.73.6.9.

Chapter 11
A Socioscientific Issues Approach to Environmental Education

Benjamin C. Herman, Troy D. Sadler, Dana L. Zeidler, and Mark H. Newton

Abstract This chapter makes a case for applying a socioscientific issues (SSI) approach to environmental education (EE). We present a model for SSI-based teaching and learning that can be used for the development and implementation of EE learning experiences. The model highlights the significance of design elements, teacher attributes, learner experiences, classroom environment, and peripheral influences. We present a description of and results from an experiential environmental issues course as a means of showcasing an implementation example of the model for SSI-based for teaching and learning. The course features contentious environmental issues from the Greater Yellowstone Area with a particular focus on the reintroduction of wolves. The chapter also presents evidence related to how framing this EE course with an SSI approach led to student development of competencies including conceptualizing scientific claims, balancing ethical and cultural considerations, negotiating unintended consequences of proposed solutions, and engaging in socioscientific discourses.

B.C. Herman (✉)
Department of Learning Teaching & Curriculum, College of Education, University of Missouri, Columbia, MO 65211, USA
e-mail: hermanb@missouri.edu

T.D. Sadler
School of Education, University of North Carolina at Greensboro, Greensboro, NC 27412, USA
e-mail: tdsadler@uncg.edu

D.L. Zeidler
Department of Teaching and Learning, College of Education EDU105, University of South Florida, Tampa, FL 33620–5650, USA
e-mail: zeidler@usf.edu

M.H. Newton
Department of Science Education, College of Natural Sciences, California State University, Chico, CA 95929, USA
e-mail: mhnewton@csuchico.edu

G. Reis, J. Scott (eds.), *International Perspectives on the Theory and Practice of Environmental Education: A Reader*, Environmental Discourses in Science Education 3, https://doi.org/10.1007/978-3-319-67732-3_11

11.1 Environmental Education and Socioscientific Issues

There are numerous possible approaches to environmental education (EE). Teaching materials and learning exercises might project a strong environmental advocacy perspective, while other approaches might foreground scientific principles and processes (Duerden and Witt 2010). The science-first approach often assumes that if students are armed with scientific knowledge, then they will enact pro-environment decisions and behaviors (Ballantyne and Packer 1996). Other EE approaches emphasize affective (e.g., Pooley and O' Connor 2000), cultural, and social justice (e.g., Cole 2007) dimensions to promote pro-environmental dispositions.

Some learning experiences may focus on environmental health at a global level. These are congruent with the themes of "thinking globally" advocated by EE initiatives such as the UNESCO–UNEP International Environmental Education Programme (Gough 2013). Other approaches drive instruction through the exploration of a specific, local environmental problem and its impact on indigenous cultures. For instance, Anna Cole (2007) reflects about how her place-based EE curricula could have facilitated students to consider environmental and social justice topics such as the perspectives of various stakeholders (e.g., Native Americans, Hispanic ranchers) impacted by river ecosystem issues in northern New Mexico. Thus, EE may create opportunities for learners to connect personally with issues through field-based or place-based experiences, while others are more abstracted and presented in generalized terms (Kurdyavtsev et al. 2012). Whether intentional or tacit, pedagogical decisions carry with them a number of assumptions related to what ought to be valued, how people learn, what the outcomes of education should be, and how learners best engage with their environment.

For the past 15 years, members of our authorship team have collaborated through research, curriculum development, teacher education programs, and implementation efforts in authentic settings to develop, empirically explore, and theoretically justify the socioscientific issues (SSI) framework (Zeidler 2014). We see the SSI framework as a useful perspective and lens for approaching teaching and learning challenges in multiple contexts including EE (with science education and integrated science, technology, engineering, and math [STEM] education serving as other examples).

SSI represent ill-structured problems at the intersections of science and other aspects of society. They tend to be controversial; multi-faceted; subject to multiple, sometimes, contradictory perspectives; and connected to scientific concepts. Despite the necessary association between SSI and underlying science ideas, solutions for SSI are underdetermined by scientific data alone. SSI encompass a wide range of real-world issues including contentious environmental issues (CEI) such as climate change, hydrofracturing, and the introduction (or reintroduction) of flora and fauna into natural communities. SSI-based education leverages the compelling ethical nature of these issues, the significance of decisions regarding these issues, and the chance to connect learning opportunities to the lived experiences of students. In terms of commitments most relevant to EE, the SSI framework presumes the following:

- Educating citizens capable of responsible citizenry (including environmental stewardship) requires development of conscience. By conscience, we refer to one's self-awareness, self-regulation, and explicit moral recognition of being an impactful component of a larger system.
- CEI (e.g., climate change, hydrofracturing, species reintroduction) are inherently challenging and cannot be solved through simple means.
- Learners require opportunities to explore complex problems, negotiate multiple solutions, and develop and justify their own perspectives.
- Understanding the science underlying CEI is necessary but not sufficient for resolving these problems.
- Opportunities to explicitly negotiate multiple dimensions of CEI (e.g., unequal impacts on diverse groups and the environment, ethical concerns, political and ideological dimensions) are critical for learning and fostering responsible scientific literacy, citizenship and environmental stewardship.

In summary, we propose that SSI-based education provides an ideal pathway to help people conceptualize and respond to the various facets (e.g., scientific, social, political, and ethical) CEI entail. SSI approaches encourage students to reflectively consider multiple perspectives, from personal to global, when weighing the unequal positive and negative trade-offs that accompany CEI resolution. SSI teaching and learning can promote civic responsibility and scientific literacy necessary for environmental and sociocultural stewardship.

11.2 A Model for SSI-Based Teaching and Learning

Over the last few years, we have developed a model to help translate the theoretical vision along with existing empirical evidence for SSI teaching and learning. The model first emerged through the analysis of nine successful instantiations of SSI-based teaching from around the world including some that focused on environmental issues, healthcare challenges, biotechnology issues, and genetic engineering (Sadler 2011). We presented a revised version of the model as a tool for curriculum designers, teachers, and school administrators interested in using SSI in school settings (Presley et al. 2013) and more recently used the model as the basis for a conceptual framework for aligning EE with science education (Sadler and Murakami 2014). In this chapter, we present the model for SSI-based teaching and learning along with a description of a hybrid classroom and informal EE course showcasing its implementation.

A graphic representation of the model for SSI-based teaching and learning is presented in Fig. 11.1. Interpretation of the model begins with the core aspects, which are situated most centrally: design elements, learning experiences, and teacher attributes. While it is possible to distinguish among the core elements for

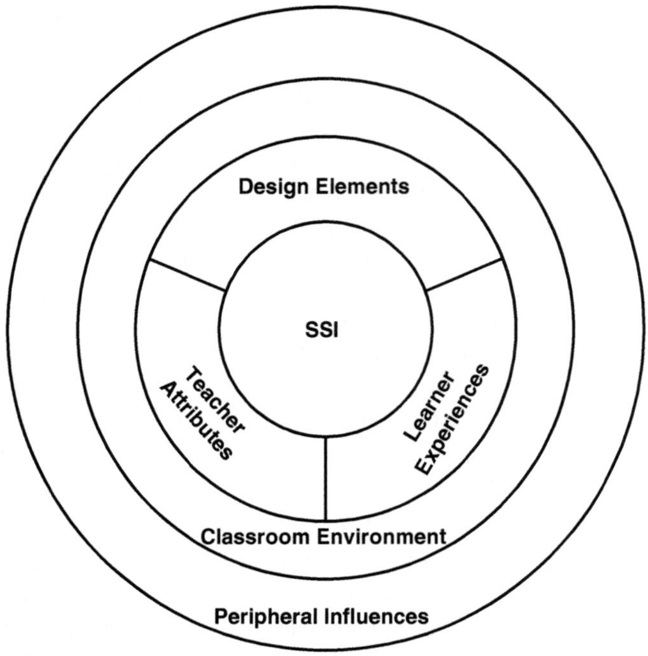

Fig. 11.1 Graphic representation of the model for SSI-based teaching and learning (Presley et al. 2013)

organizational purposes, they are interrelated. Design elements refer to issues and opportunities that should be incorporated in the design, development, and implementation of SSI learning experiences. The model prescribes four specific elements of design that need to be included in SSI-based teaching and learning. First, instruction should be built around a compelling issue; that is, the issue should be the central organizing feature for instructional planning and implementation. Other approaches to environmental education may prioritize other organizing elements for the design of instruction such as standards or content. In the case of SSI-based teaching, instruction may be aligned with standards and address important content, but the decisions around which standards and content to address are informed by the issue to be explored. This prioritization of the issue helps to ensure that the ideas addressed are meaningful and contextualized with respect to the issue being considered.

Second, in the implementation of SSI-based teaching, the issue should be presented early in the experience. Many approaches to science and EE use issues as examples of phenomena or principles, and these examples may be presented at any point during instruction. In order to foreground the focal issue within SSI-based teaching and learning, the issue itself should be presented very early within the teaching sequence.

Third, SSI instruction should provide scaffolding for student engagement in higher-order practices. These higher-order practices can be framed in several ways including argumentation, decision-making, critical thinking, or socially responsible action. Student engagement and learning with respect to at least some of these (or related) practices is a fundamental goal for SSI-based teaching and learning.

Finally, design of SSI learning experiences should incorporate a culminating experience during which learners can synthesize ideas, compare perspectives, integrate their own commitments, and use the higher-order practices featured in the learning experiences.

Learner experiences comprise the second core aspect of the SSI-based teaching and learning model. The model suggests that learners should have opportunities to engage in six different but interrelated experiences that necessarily overlap with the design elements just discussed. Within the context of SSI instruction, students should have opportunities to:

- Engage in higher-order thinking practices
- Confront scientific ideas and theories related to the issue under consideration
- Collect and/or analyze scientific data related to the issue
- Negotiate social dimensions of the issue
- Confront ethical dimensions of the issue
- Consider nature of science themes associated with the issue

Considering which of these learner experiences are afforded within a particular SSI unit of instruction can help to direct attention to areas of the instruction or curriculum materials that can be expanded and improved.

The final core aspect, teacher attributes, directs attention to teacher knowledge and characteristics. In considering the attributes possessed by teachers, who implement SSI-based instruction successfully, four commonalities consistently emerge. First, successful teachers are knowledgeable about the science content related to the issue. Supporting student exploration of the science involved in SSI requires that teachers have a reasonable degree of expertise surrounding that science.

Second, teachers also need to be aware of social considerations associated with the issue. Given the wide range of social knowledge pertinent to most SSI, it is unreasonable (and unnecessary) to expect that a teacher can have expertise in all relevant areas; however, it is very important that teachers are familiar with at least some of these ideas.

Third, SSI, by definition, are complex problems for which some information is not available. Teachers cannot know everything there is to know about a particular SSI, and they should be honest with their learners about these knowledge limitations.

Finally, given the uncertain status of SSI, and the goal of SSI-based teaching to foster community inquiry and negotiation, teachers need to position themselves as knowledge contributors within their classroom communities as opposed to the sole authority. For some teachers, adopting a role on more equivalent footing with their students can be uncomfortable, but the sharing of power and discourse within the learning environment is essential for successful SSI-based teaching and learning.

The remaining elements of the model, classroom environment and peripheral influences, highlight the point that design and enactment of learning experiences occur within contexts that significantly shape the trajectories of engagement and learning. Classroom environment is positioned as the first and most temporally relevant source of influence. Dimensions of the classroom environment such as expectations for participation, discourse norms, relationships among members of the classroom community, and resources available will play huge roles in determining how curricular and pedagogical plans are enacted and the kinds of outcomes that will be achieved.

Peripheral influences are positioned immediately beyond the classroom environment. In actuality, this single dimension of the model represents a number of possible spheres of influence that may shape the classroom environment including teachers and students. The classroom environment is shaped by the institutional environment, which, in turn, is necessarily influenced by the community in which the institution is situated. National trends and standards, economic forces, and broader expectations are some of the many peripheral influences that can ultimately affect ways in which SSI-based EE unfolds. However, the point of the model is not to suggest that educators identify all of the possible peripheral influences but rather to develop an appreciation for and sensitivity to the contextual realities of teaching through student engagement in complex and controversial issues.

11.3 A Case of Using the SSI Model for Environmental Education: Contentious Environmental Issues in the Greater Yellowstone Area

In this section, we pivot from the conceptual presentation of the SSI-based teaching and learning model to an applied case profiling an SSI approach for EE. Presented here is an SSI-embedded experiential environmental issues course co-taught by several of the authors that focused on Greater Yellowstone Area (GYA) CEI (e.g., reintroduction of wolves, brucellosis in bison, ranching practices). A notable feature of this course was the extent that the 36 enrolled undergraduate students from diverse majors (e.g., nursing, law enforcement, art, music, education) were immersed for 10 days in authentic experiences with GYA CEI. The students then completed a 3-week online component after returning to their university where they developed a written analysis of a CEI. We present a description of the course and its impacts on student learning; the presentation is organized around the four specific design elements highlighted in the SSI-based teaching and learning model.

Table 11.1 Sequence of interrelated events and the issues addressed

Portion of course	Course activity
Predeparture	Discuss the analogy that the university is to the town (Farmville, VA) as Yellowstone National Park (YNP) is to the GYA
	Interview citizens of Farmville about their feelings about the university and the students. Group discussion on findings (e.g., perspectives, biases) and make connections to the GYA
	Introduce the wolf reintroduction and management issue in the GYA
GYA field experiences	Interact with Jackson Hole residents and Wyoming fish and game biologists about GYA CEI
	Travel through YNP with field stops at key areas including visitors' centers and natural features
	View wildlife in YNP. Interact with wolf ecologists, tour guides, and local residents
	Interact with ranchers possessing progressive and traditional perspectives about GYA CEI. Interact/discuss with Gardiner residents about GYA CEI
	Hike YNP with a naturalist. Interactive presentation with a nature activist/writer
	Students placed into groups based on different perspectives modeled in GYA to prepare arguments for stakeholder council meeting assuming perspectives encountered in GYA and discuss possible resolutions of wolf reintroduction/hunting issues
	Travel through YNP with field stops at nature areas and interactions with Native Americans about GYA CEI
	Students conduct community stakeholder meeting and develop proposed resolutions about wolf issue
	Participate in culminating town hall-style forum on wolf hunting quotas in Montana. Students assume perspectives of GYA stakeholder encountered during experiential field component of course
Post-fieldwork (online)	Analyze public documents and complete a major writing assignment on natural resources management issues

11.3.1 Design Element 1: Build Instruction Around a Compelling Issue

At the forefront of SSI instruction is a compelling and contentious issue. Table 11.1 outlines the sequence of events and the compelling issues addressed during the CEI course. Important to note is that our design of the course was fluidly responsive to the learners' experiences. For instance, the instructors would facilitate the students' learning experiences (e.g., presentations from those impacted by CEI, visits to nature centers) and help them deconstruct those experiences through reflexive pedagogical moves (e.g., modeling questioning practices during presentations and then later helping students understand the intent of the questions and the information they solicited). Furthermore, course experiences were deliberately sequenced so the instructors could scaffold students to autonomously consider multiple perspectives and the scientific, ethical, and sociocultural dimensions related to CEI.

Reflective of the complexities associated with CEI, the course instruction employed an ecological approach by stressing that many environmental issues are interrelated, have broad reaching impacts on humans and nature, and require consideration of many obvious and nuanced perspectives. Several interrelated environmental issues exist in the GYA and were addressed throughout the course. For instance, stakeholders such as ranchers, hunter outfitters, wildlife biologists, nature advocacy groups, and Native Americans possess different perspectives and debate about how to resolve CEI such as elk harvest quotas, bison home ranges and brucellosis, and sustainability of ranching practices. All of these issues are controversial, and the arguments from the stakeholders draw from multiple perspectives including those that are scientific, ecological, sociocultural, ethical, and economic in nature.

At the forefront of the CEI addressed throughout the course was the reintroduction and management of wolves in Yellowstone National Park. According to many GYA residents, wolf reintroduction is the most contentious and polarizing environmental issue in that area. Matthew Wilson (1997) notes that the wolf reintroduction issue is the symbolic representation of the contention between those whose interests are environmental preservation and those whose interests are the economic utility of the Greater Yellowstone Ecosystem. Furthermore, Wilson (1997) points out that unequal access to social power, conflicting beliefs about property use, and discrepant perceptions about nature fuels the controversy surrounding wolf reintroduction. Despite the contention that pervades this issue, scientific evidence has established that wolves are an important member of the GYA ecosystem justifying their restoration to sustainable population levels. This issue is at the heart of the CEI course and remains relevant and controversial (Smith and Bangs 2009).

11.3.2 Design Element 2: Present the Issue Early in the Experience

The contention surrounding wolf reintroduction in Yellowstone National Park was introduced early in the course prior to the students arriving in the GYA. In line with the scaffolding approach advocated within the SSI instructional framework, we utilized local CEI (e.g., impacts of their university's expansion on the local natural and built communities) familiar to the students helping them conceptualize how to engage SSI through argumentation and evaluating multiple perspectives, scientific evidence, ethical considerations, and sociocultural factors. We then introduced students to GYA CEI by focusing on wolf reintroduction and asking them to consider ways in which resolving SSI requires varied approaches. For instance, we asked the students to contemplate how multiple perspectives and environmental ethical considerations might influence how the university expansion is resolved and how that might be similar to underlying issues influencing the wolf reintroduction dispute. Our rational for such an approach is that we recognized that the students' familiarity with locally relevant CEI could be used as a springboard to facilitate their engagement with the less familiar GYA environment and CEI.

11.3.3 Design Element 3: Provide Scaffolding for Higher-Order Practices

Wilson (1997) writes about the contested nature of the GYA with on the one hand its resonating and prolific natural beauty and on the other the power struggles that accompany land ownership and natural resources management. Specifically, the federal government owns over half of the GYA county land base, and thus all policy decisions such as wolf reintroduction profoundly impact local residents and thus spur controversy around these issues. As Fig. 11.1 illustrates, peripheral factors and the learning environment must be considered when designing SSI instruction. The experiential field component of the SSI instruction profiled here was crucial for immersing students in GYA CEI, so they better understood the nuanced contentious factors associated with wolf reintroduction and management. Throughout the span of the experiential field component, our objective was to scaffold students from a passive role through witnessing the instructors and stakeholders interacting and deliberating relevant SSI themes to becoming active participants who interacted with and took ownership in learning about the stakeholders' backgrounds, cultures, and concerns.

To gain an initial sense of the sociocultural aspects of GYA CEI, the students began in Jackson, Wyoming, where they spent 3 days investigating how various stakeholders perceive CEI including wolf reintroduction. For example, the students spent 1 day talking with residents about their views on wolf reintroduction and how those views were developed. The students also experienced a presentation by area wildlife biologists focusing on the integrated nature of CEI in the GYA such as how various groups perceive wolf management in relation to elk harvests. During early interactions with stakeholders, the authors modeled questioning strategies for the students that reflected many of the subtle aspects needed to investigate socioscientific issues including skepticism, consideration of multiple perspectives, sociocultural factors, ethics, and scientific evidence. The following dialogue provides one example between the instructors and the biologists exemplifying this process:

Instructor Can you talk about the reaction to some of your management practices?

Biologist We have a really diverse public here. We have everything from very traditional 3rd generation Wyoming cowboy hunter type to very environmental green and everything in between.... We have a lot of different views and backgrounds.

Instructor You talked about wolves. In a group I was in today ... a lot of newspapers and magazines we found addressed the recent issue of how the wolf hunting regulations have been redone and the state is now in charge. That was the headline story in two or three of those.

Biologist Wolves were reintroduced to Yellowstone… Until recently they were on endangered lists. They were managed by the federal government. Through lots of politics and court rulings they eventually came off the endangered lists and are under state management for just over 2 years now. So, since they have become under state management and off the endangered species list, we manage them like any other trophy game species: black bears, mountain lions, etc. so people hunt them.

Instructor What do you do with the diversity with the different kinds of groups? You mentioned in the beginning with the different agencies and how they work together in some way… So, can you give me a sense of how these diverse groups, each with vested interests in their own careers, come to consensus to work together? You say it is politics. Can you give me an extended sense of how you negotiate from such diverse points of view?

Biologist There have been times when the agencies have not seen eye to eye and have not gotten along. They went through bad times. I think there is general agreement hopefully at this point from all the different players. We can accomplish more if we do compromise on the things we can compromise on. There are things our agency says, that we cannot compromise. Sorry we just can't work together on this issue. Fundamentally then science is not speaking for itself. You are just polishing it. You try and figure and get a collection and distribute it in a way where you do have an optimal solution, but in a way where everybody gets something. I think there is after years of doing that, there is a tradeoff of like … we are going to compromise on this but remember that the next issue comes up … maybe you guys need to compromise. You have years and years of relationships going on and you have a trust in a way. It comes back down to a lot of people dealing with people. I wasn't really trained to deal with people. Ha ha. It's a lot of dealing with people.

Instructor It seems like when you are dealing within your own agency that some of the discussions go beyond the scientific evidence. You can see we can sustain and manage … to what extent do you sometimes discuss other considerations… such as the ethics of the situation or how it will affect the socioeconomics of the region?

Biologist We actually put a lot of things into the pot when we make a management decision. We try to get as much scientific evidence as we can and that is the base. We have to then make sure we hear all publics concerns. Lots of meetings, etc. The wildlife is held in the public trust. We have public commissions. We have a group of six commissioners, which are just regular citizens that are appointed to the commission board. They are the ones that decide what really they should do. We build the framework and then the public decides.

Instructor To what extent have you had an actual example where you had an ethical consideration override the science evidence? For instance, some groups that have more of a spiritually vested interest in the management that may think it is ethically wrong on how you are managing wildlife populations. Or even a socioeconomic consideration overrode your best judgment from the scientific evidence.

After such interactions, the instructors would engage in reflective discussions with students in formal (whole class meetings) and informal (car rides to field experiences) settings. These discussions were crucial for scaffolding students to consider and engage in discourse about the more complex nuances of GYA CEI.

Over the next 7 days, students transitioned from Jackson to Gardiner, Montana, and Yellowstone National Park (YNP) where they critically examined CEI with a focus on wolf management. Specifically, the students met with two local ranchers with diverse perspectives regarding this issue. The first held progressive ideas on how to protect her animals from wolves through hazing and using range riders to manage herds. The second rancher held more traditional beliefs on protecting his stock, such as trapping and shooting wolves that ventured onto his property. The students also accompanied a wolf ecologist and a naturalist who charter tours into YNP to view wildlife and discuss wolf reintroduction. The students concluded scheduled stakeholder interactions by gaining the perspectives of an environmental activist and author and a member of the Crow Native American tribe. In addition to these formal interactions, students were required to engage with local stakeholders throughout the communities (e.g., at restaurants, at gas stations, and at the hotel) they visited. The explicit modeling from the instructors resulted in students beginning to engage in sophisticated discourse about CEI with one another and stakeholders, suggesting they were considering more complex facets of SSI resolution. Two examples below are questions asked by students to wolf ecologists and naturalists that exemplify how they were conceptualizing CEI from increasingly sophisticated vantage points: "Do you think it might be the negative cultural image of the wolf that it becomes the scapegoat for other perceived natural resources problems?" "How do you think the fact that people hunt elk here affects their views on wolf hunting?" "Do you feel there is a positive relationship between Native Americans in the area and others that live here despite past tensions and how the park is now managed as public land?"

11.3.4 Design Element 4: Culminating Experience

The culminating experience was a town hall-style forum on wolf management in the GYA. Three days prior to the forum, the students were assigned to one of six groups representing a GYA stakeholder and asked to develop solutions and accompanying justifications (see Table 11.2). Important to note, the instructors deliberately assigned groups that held diverse and, at times, polarized views about wolf management to emphasize the contention surrounding this issue and instructed students to

Table 11.2 Town hall forum stakeholder groups' proposed solutions and justifications regarding wolf management

Group	Summary of proposed solution and justification
State wildlife managers	Wolves should be sustainably harvested through controlled hunts. Natural resources and YNP are for the benefit and enjoyment of the people. Social, economic, and ecological considerations must be weighed. Wolves should be present to maintain ecosystem balance and tourism. However, wolf numbers should be limited to mitigate economic losses (e.g., livestock predation) and negative environmental consequences (diminished elk populations)
Traditional ranchers	Wolves should be heavily hunted and state managed, not federally. Ranching is a long-standing family tradition that wolves threaten by predating on cattle which can cost thousands. This threat combined with other economic hardships associated with ranching threatens to end the ranching culture that has been built through many generations
Progressive ranchers	A quota of 20 to 25 per district is feasible but wolves should only be killed when necessary (e.g., extreme livestock predation). The culture of ranching must change. Wolf reintroduction has occurred, and coexistence between ranchers and wolves is possible through new ranching and wolf-deterring methods (e.g., range riders, hazing, electric fences with flaggery, removal of sick or dead livestock and elk)
Ecologists/wildlife viewing tour guides	Wolf harvests outside of park could remain at current levels. However, wolf hunting should be restricted within YNP and a buffer zone around the park that will be determined every 10 years based on park packs' home ranges. Wolves were an important component of the GYA ecosystem when the park opened in 1872 but have only recently recovered from a cruel extirpation by humans. Killing one wolf can break up packs and orphan pups. Wolves can sustainably manage themselves within park boundaries and could be tracked through radio collars. Typically, park packs stay within YNP 98% of the time, and only 1% of cattle deaths outside of the park are confirmed wolf kills. Furthermore, the concern of wolves overkilling elk is unjustified. The primary killers of elk are humans and winterkill, not wolves. Unfortunately, the media portrays wolves negatively, but they are actually very important economically and ecologically
Environmental activists	The killing of wolves should be completely prohibited. Wolf protection buffer zones around the park are irrelevant. Wolves have an intrinsic right to be here just like humans and nature's other creatures. Nature should be left alone to take its course. Wolves, bison, elk, and other species were part of the ecosystem balance long before humans' presence. Humans have eliminated other species such as cutthroat trout, and without wolves, elk and deer populations in the park would be unregulated. This would cause negative environmental impacts such as over grazing and browsing of vegetation

(continued)

Table 11.2 (continued)

Group	Summary of proposed solution and justification
Outfitters and hunters	Montana state annual wolf harvest limits should be increased from 230 to 300 and the limit of five wolves per person eliminated. Wolves can be preserved in YNP but protection buffer zones are unjustified. Hunting elk is a cultural tradition among families that is also important to their economic well-being and livelihood. Wolves negatively impact elk numbers. Hunters bring outside money to the economy, and increasing the wolf limit will benefit ranchers. Furthermore, there is less chance the Fish and Game Department will need to reimburse ranchers for livestock losses, and the money for wolf licenses can go toward wolf preservation in the park. Adjustable buffer zones are unjustified given YNP is 2.2 million acres and buffer zones will constantly expand with wolf dispersal

make a concerted effort to represent the interests of their group rather than those of their own perspectives. Within these groups the students developed a proposal about wolf management including wolf hunting limits and protective buffer zones around YNP. Three students were selected to serve as a federal panel that evaluated the stakeholder group's arguments and given the charge of issuing a detailed ruling on wolf management. Prior to the forum, all of the groups met with instructors to discuss issues associated with using and evaluating different forms of evidence and criteria for decision-making.

The 2-h town hall forum took place at a local restaurant and tavern – an authentic setting where residents of the area typically exchange ideas about local CEI. The forum was student conducted with the instructors assuming a facilitator role. Each stakeholder group prepared a 5-min opening statement presenting their resolution for wolf management and harvest limits to a mock federal panel (in this case, the panel consisted of three students selected by instructors). After the opening statement, the federal panel asked each group clarifying questions. Upon completion of the opening statements and panel questions, each group was provided time to prepare and orate rebuttals to opposing groups. The rebuttals were followed by closing arguments from each group that summarized their proposed resolution and supporting evidence. The federal panel was then sequestered to deliberate and reach a decision with accompanying justifications about how to manage wolves including an annual wolf harvest limit.

The federal panel evaluated arguments and determined a ruling based on five preestablished criteria: (1) pragmatism (best solution for all groups), (2) robustness (resiliency and durability of the arguments), (3) evidence based (scientific, logical, moral, and cultural truths), (4) feasibility, and (5) precedence (e.g., based on historical events and prior policies). The panel's ruling included a wolf-protective buffer zone around YNP with continued research on wolf population distributions and buffer zone adjustments every 10 years. Furthermore, the panel decision established a wolf hunting limit up to 20% of the state population annually, and that wildlife agencies would assist ranchers' predator deterrence efforts and elimination or relocation of confirmed livestock predators.

At the conclusion of the town hall forum, the instructors debriefed the students on the culminating activity and how their course experiences enabled them to contemplate and engage in discourse about wolf management. For instance, the instructors explicitly addressed how the students were considering and discussing multiple perspectives, ethics, scientific evidence, and sociocultural factors when engaging SSI. Furthermore, the instructors stressed the importance of assuming an eco-justice perspective personally, civically, and globally needed to resolve CEI through proenvironmental behaviors.

The final 3 weeks of the course were conducted online after the students returned to their university where they developed a written analysis of a CEI. Students were expected to synthesize the information that was garnered during the field component to create a position paper discussing how to manage their specific resource issue.

11.4 Learner Considerations and Impacts of the Course

Throughout the previous sections, we have attempted to explicate how our design of the course was fluidly responsive to the learners' experiences and conceptions – existing and newly acquired through the CEI course. The course experiences were deliberately sequenced so the instructors could scaffold students to autonomously consider nuanced factors (e.g., scientific, ethical, and sociocultural dimensions) of CEI resolution. During these interactions and subsequent discussions with instructors, the students learned important lessons regarding SSI engagement such as how to conceptualize scientific claims (e.g., accuracy and the nature of those claims) in juxtaposition with ethical and cultural considerations and possible unintended consequences of CEI resolution. As a result of these experiences, the students began to take ownership of their engagement with GYA CEI through actions such as eliciting perspectives from stakeholders through sophisticated questioning and debating with peers and the course instructors about those perspectives. Evident among student discourse was the realization that science is among many valuable ways of knowing that must be judiciously weighed when engaging SSI and that sociocultural, ethical, and economic consequences typically accompany CEI resolution. For instance, scaffolding interactive experiences with a variety of stakeholders, from wildlife biologists and ranchers to Native Americans and activists, who consistently deal with GYA CEI, modeled for students how to draw from scientific evidence as well as indigenous and traditional ecological knowledge, values, and beliefs. We viewed these experiences as crucial in helping many students to move from a position resembling scientism where it is perceived that CEI should only be resolved through scientific truths, engineering, and techno-centered approaches to a more rational and balanced view that weighing additional factors (e.g., cultural, ethical, and historical) must occur when engaging CEI. The following student's sentiments before and after the course exemplify this conceptual shift:

> With advancing scientific research they are gaining more proof about the negative effects that humans have on the environment. You cannot argue with the fact that species are becoming extinct, and glaciers are melting. (Samantha's pre–course views)

> I have learned that scientific evidence is a good base but other things such as morals, culture, and ethics also come to play when making decisions. Scientific evidence can also be bias in the way it is presents and what part of the research is publicized. Many things are complicated, and the scientific reasoning may change, it is not set in stone. I think that science will help with natural resource issues, but I do not think it is the end–all–be–all. (Samantha's post–course views)

Reflective of the course focus, many students drew from their Yellowstone experiences to conceptualize the complexities associated with CEI in their home communities. For instance, one student reflectively linked her Yellowstone experiences with issues her home community is facing by stating:

> I noticed after I went to Yellowstone, I did not consider the different perspectives of issues as much as I had thought. Now I try to always take different perspectives on natural recourse management issues. For example, in my county there is a suggestion to make Powhatan's home a national park. When I heard this after I came back from Yellowstone my initial thought was, what do the nearby locals of that area think vs. the people who are working to get the park in motion? I continued to think of the county locals in various areas and jobs and how they might perceive this issue as well as the impact it might have on the land and wildlife in that area. I think that natural resources management issues can raise social and ethical concerns and conflicts...(Valerie's post–course views)

11.5 Teacher Attributes and Recommendations

The SSI teaching and learning model highlights teacher attributes as a core component of successful SSI instruction. The instructors of the GYA CEI course possessed advanced degrees and extensive professional experiences in wildlife biology, geology, and science education. They also worked extensively in the GYA conducting ecological research and educational outreach. Therefore, they were familiar with many of the scientific and social considerations associated with GYA CEI addressed; however, they also recognized their limitations regarding these knowledge domains and worked with students to create a community of inquiry where knowledge about how to engage CEI such as wolf management was co-constructed through "real-time" experiences and discussions about those experiences. Thus, instead of assuming an authoritarian role, the instructors consistently modified instruction based on knowledge they learned with the students.

Using issues to frame EE is an intuitively appealing approach, and while an issue-oriented approach limits the wide range of possibilities for EE, there are still many critical decisions that must be made by designers and instructors using environmental issues. We have found the SSI framework to provide a powerful model for informing those decisions in ways that ensure alignment between our theoretic commitments, pedagogical priorities, and educational objectives. The model presented within this chapter specifies the SSI approach with a level of detail that can help to inform design decisions and implementation choices, and the GYA CEI course provides an example of how this model played out in an actual environmental education context. From our perspective, productive advancements to EE can be made by more widespread applications of the model for SSI-based teaching and learning accompanied by rigorous testing of various dimensions of the model. We cannot stress enough how such progressive pedagogical approaches are crucial for effectively implementing SSI and promoting the kind of scientific literacy and sociopolitical action necessary for civil democracy and environmental sustainability.

Questions

Discussion questions that could help instructors and students to engage in meaningful conversation about the ideas presented in this chapter.

1. Consider various local and global environmental issues. In what ways do these exhibit characteristics that would make them socioscientific issues (SSI)?
2. How does an SSI instructional approach foster scientific literacy in a manner that enables people to engage in citizenship and democratic decision-making?
3. In what ways does SSI instruction promote multiple perspectives (e.g. scientific, sociocultural, ethical) regarding contentious environmental issues resolution?
4. How must teachers be prepared so they are able to effectively incorporate SSI instruction in environmental education courses?
5. Consider various local and global contentious environmental issues. How would you implement an SSI approach to promote student engagement with those issues?

References

Ballantyne, R., & Packer, J. M. (1996). Teaching and learning in environmental education: Developing environmental conceptions. *Journal of Environmental Education, 27*(2), 25–32.

Cole, A. G. (2007). Environmental education principles through multidisciplinary frameworks. *The Journal of Environmental Education, 38*(2), 35–44.

Duerden, M. D., & Witt, P. A. (2010). The impact of direct and indirect experiences on the development of environmental knowledge, attitudes, and behavior. *Journal of Environmental Psychology, 30*(4), 379–392.

Gough, N. (2013). Thinking globally in environmental education: A critical history. In R. B. Stevenson, M. Brody, J. Dillon, & A. E. J. Wals (Eds.), *International handbook of research on environmental education* (pp. 33–44). New York: Routledge.

Kurdryavstev, A., Stedman, R. C., & Krasny, M. (2012). Sense of place in environmental education. *Environmental Education Research, 18*(2), 229–250.

Pooley, J. A., & O'Connor, M. (2000). Environmental education and attitudes: Emotions and beliefs are what is needed. *Environment and Behavior, 32*(5), 711–723.

Presley, M. L., Sickel, A. J., Muslu, N., Merle-Johnson, D., Witzig, S. B., Izci, K., & Sadler, T. D. (2013). A framework for socio-scientific issues based education. *Science Educator, 22*(1), 26–32.

Sadler, T. D. (2011). Socio–scientific issues based education: What we know about science education in the context of SSI. In T. D. Sadler (Ed.), *Socio–scientific issues in the classroom: Teaching, learning and research* (pp. 353–367). New York: Springer.

Sadler, T. D., & Murakami, C. D. (2014). Socio–scientific issues based teaching and learning: Hydrofracturing as an illustrative context of a framework for implementation and research. *Brazilian Journal of Research in Science Education, 14*(2), 331–342.

Smith, D. W., & Bangs, E. E. (2009). Reintroduction of wolves to Yellowstone National Park: History, values and ecosystem restoration. In M. W. Hayward & M. J. Somers (Eds.), *Reintroduction of top–order predators* (pp. 92–125). Hoboken: Wiley–Blackwell.

Wilson, M. A. (1997). The wolf in Yellowstone: Science, symbol, or politics? Deconstructing the conflict between environmentalism and wise use. *Society and Natural Resources, 10*(5), 453–468.

Zeidler, D. L. (2014). Socioscientific issues as a curriculum emphasis: Theory, research and practice. In N. G. Lederman & S. K. Abell (Eds.), *Handbook of research on science education, volume II* (pp. 697–726). New York: Routledge.

Chapter 12
Moving Forward from the Margins: Education for Sustainability in Australian Early Childhood Contexts

Sue Elliott and Julie Davis

Abstract The early childhood years are regarded as the most significant period for human development. Yet, education investments in environmental education (EE)/ education for sustainability (EfS) have been largely overlooked in the early years. This chapter offers an analysis of early childhood education for sustainability (ECEfS) theory and practice in Australia – recognised as a leader in ECEfS. It draws on historical and international contexts, key theoretical perspectives and policy directions informing broader uptake. We believe the Australian experience with ECEfS offers pointers for other parts of the world looking to ramp up their ECEfS activities.

12.1 ECEfS in Context: Moving Forward from the Margins Historically and Internationally

Until recently, early education investments have been largely ignored internationally, and there remain many majority world regions where there is still no systemic access to preschool education or where the quality is poor (OECD 2012). However, an expanding body of literature across economics, sociology and health indicates that early investments in human capital offer substantial returns for individuals and their communities with long-term positive impacts for the future. For example, economic modelling has identified that for every dollar allocated to early childhood education, over 15 dollars are saved in remedial interventions during later childhood and young

S. Elliott (✉)
School of Education, University of New England, Armidale, NSW 2351, Australia
e-mail: sue.elliott@une.edu.au

J. Davis
Faculty of Education, Queensland University of Technology,
2 George Street Brisbane, Brisbane, QLD 4000, Australia
e-mail: j.davis@qut.edu.au

G. Reis, J. Scott (eds.), *International Perspectives on the Theory and Practice of Environmental Education: A Reader*, Environmental Discourses in Science Education 3, https://doi.org/10.1007/978-3-319-67732-3_12

adulthood (Calman and Tarr-Whelan 2005). Such modelling is supported by brain research (Shonkoff and Richter 2013), which confirms through brain imaging just how critical early experiences are to optimal brain development. The early childhood years are unequivocally the most significant period of human development.

In response to this economic modelling data and spurred by the "Starting strong 11: Early childhood education and care" report (OECD 2006), the Australian government began implementing policy directives to improve early childhood education (COAG 2008). In parallel with these policy directives, the early childhood years have also been recognised as a highly pertinent life phase for integrating understandings about sustainability and an ethic of sustainable living through early childhood education pedagogy and practice (UNESCO 2008a). We view sustainability just as important as the integration of healthy eating and childhood physical activity and the early addressing of social justice issues including gender and race stereotyping. The UNESCO report summarising the achievements of the United Nations' Decade of Education for Sustainable Development (2014) states:

> Young children are both current and future citizens with already existing capabilities to shape sustainable societies. Investments to build their awareness, values, knowledge and capacity for sustainable development will serve to set the world on more sustainable pathways now and into the future. (p. 78)

Reflecting on the history of EE/EfS dating back to the first intergovernmental conferences (UNESCO/UNEP 1977) and the 1980s school EE initiatives (Gough 1997), early childhood ECEfS was largely overlooked both in Australia and internationally. During these earlier periods, broadly entrenched developmental paradigms arising from psychological epistemological viewpoints of learning and development, suggested young children to be incapable of understanding the abstract complexities of sustainability, let alone contributing to addressing environmental issues in any substantive way (Elliott and Davis 2009). However, such paradigms about children's capabilities were beginning to shift in the 1990s informed by the work of Allison James and Alan Prout (1990), who viewed children as active social participants, both capable and agentic. Further, the release of the "United Nations Convention on the Rights of the Child" (UNICEF 1989) promoted children's rights to participate in decision-making about matters impacting them. These new constructs significantly aligned with EfS participatory approaches, and, as Julie Davis (2007) later asserted, young children can be "problem seekers, problem solvers and action takers in their own environments" (online). In addition, the principles of EfS as espoused in the Australian "Living Sustainably: National Action Plan" (Commonwealth DEWHA 2009) clearly aligned with contemporary early childhood philosophy and pedagogy offering a "pedagogical advantage" (Elliott and Davis 2009, p. 68) for early childhood educators in ECEfS. As Ingrid Pramling Samuelsson and Yoshie Kaga (UNESCO 2008a, p. 13) state, "it is not necessary to invent 'new' pedagogies in order to 'do' education for sustainability in the early years – one can build on its pedagogical traditions to do so".

Thus, internationally, a convergence between the prioritisation of the early years and shifting paradigms about children and early childhood education created a fertile space "for moving forward from the margins". ECEfS was first highlighted internationally by UNESCO (2008b) in "The Gothenburg Recommendations on Education for Sustainable Development". The subsequent "UN Decade of Education for Sustainable Development" (DESD) (UNESCO 2014) final report has reinforced the unique role of early childhood education in creating sustainable futures for all.

Aligned with these documents, since the early 1990s, a handful of individual practitioners and professional networks in Australia have navigated through this convergence with the advocacy mantra "mainstream not marginal" (Davis 1999, p. 2). Somewhat later, the field was still typified by "patches of green" in the first national review of early childhood EE (NSW/EPA, 2003). However, the raft of Australian government early childhood education policy directives (COAG 2008) set against the backdrop of UNESCO's DESD (2005) initiatives have been instrumental in guiding systemic change across the early childhood education field. The uptake of ECEfS by practitioners in Australia is now readily visible in children's programmes, physical settings and in overall service operation.

Uptake by practitioners is also supported by an escalating international research agenda where Australian researchers are key players (Davis and Elliott 2014). Albeit, not so long ago, ECEfS research literature was described as a gaping "hole" (Davis 2009). The current emerging (but dynamic) research environment invites multiple paradigms, methodologies and theoretical frames and seeks to investigate "paradoxes, holes, barriers, silences and gaps" (Davis and Elliott 2014, p. 13). While current research draws mainly on qualitative paradigms, some studies do offer quantitative approaches (McNichol, Davis and O'Brien 2011). Indeed, quite a diverse array of theoretical frames and methodologies were identified amongst the research studies represented in the first published compilation of ECEfS research, "Research in Early Childhood Education for Sustainability: International Perspectives and Provocations" (Davis and Elliott 2014). Our experiences in compiling and analysing the research chapters reflected a "multiplicity platform" (Brooker and Edwards 2010), an inviting stage for researchers seeking to investigate ECEfS from diverse perspectives. The international research group that instigated this research compilation, "Transnational Dialogues in Early Childhood Education for Sustainability", currently offers online and face-to-face networking opportunities for further progressing research collaborations, innovation, critique and exchanges. While the theoretical frames and methodologies are potentially diverse, underpinning conceptions of sustainability tend to fall into two schools of thought: "in, about and for the environment" or the "social, economic and environmental dimensions" (Hedeflak et al. 2014). There is much yet to investigate from various perspectives including collective rights, ecocentrism, posthumanism and indigenous worldviews.

In sum, the uptake of ECEfS both by educators and researchers has been tardy, an untenable situation given now pressing global concerns. However, at last there is a deepening sense of having "moved forward from the margins" to the mainstream of both the early childhood education and education for sustainability fields. This bodes well for the health and wellbeing of young children, current and future generations and the planet as a whole.

12.2 Challenging and Exploring Paradigms

We continue this chapter with a critical analysis on the pervasive "greenness" of environmental/sustainability education research in the early childhood years that, even now, appears to be somewhat stuck in the 1960s "nature education" paradigm. In other words, much of the focus in initial early childhood EE publications (Wilson 1993) remains, in some quarters, on children learning and doing things "in" and "about" the environment through a nature lens, but not necessarily being agents "for" the environment through a more inclusive eco-sociocultural lens (Elliott 2014). Specifically, Affrica Taylor (2013) has brought this sustainability–nature nexus to the fore in her critical reflections about the romanticised, historical notions of children in nature that appear to be fuelling the resurgence of children's nature programmes evidenced by the international forest preschool movement (Knight 2013), and the US Nature-Deficit Disorder campaign (Louv 2005). Elizabeth Dickinson (2013) has ardently critiqued the latter, suggesting a misdiagnosis has occurred and that a deeper investigation of pathology is required if strong sustainability messages are to be promoted through nature play programmes. As Sue Elliott and Tracy Young (2015) suggest, "sentiments of stewardship" (p.3) when playing outdoors are insufficient to promote the critical and socially mediated reflections required for transforming ethics of sustainability in early childhood. Children are doing things "in" and "about" the environment through a nature lens, as advocated in many practitioner guides and supported by nature play research literature is simply not enough in the prescient global sustainability crisis.

Australia has not been immune to the resurgence of nature-based programmes with the "Belonging, Being and Becoming: The Early Years Learning Framework for Australia" (EYLF) (Commonwealth DEEWR 2009) and the more recent "National Quality Standard" (NQS) (ACECQA 2013) both promoting outdoor play in natural settings almost in lieu of more multidimensional ECEfS approaches (NSW/ECEEN and NSW/OEH 2012). However, as social constructionism, critical theory and systems theory underpin these curriculum frameworks rather than the earlier developmental paradigms, we suggest broader eco-sociocultural- and eco-centric right-based approaches are feasible in practice (Davis 2015). Within this more contemporary theoretical milieu, children are seen as capable, agentic and visible participants in addressing all dimensions of sustainability (UNESCO 2010), thus, not limited by their developmental stage or age, which implies that playing in nature will suffice in responding to sustainability issues. Examples of such broader theoretical approaches are now more evident in research studies (e.g. Engdahl and Arlemalm-Hagser 2014). Our analysis here is highly pertinent for researchers and practitioners alike, as we continue to question how we might negotiate more contemporary ideas of the sustainability–nature nexus. Otherwise, as Eva Arlemalm-Hagser and Anette Sandberg (2011) found in their investigation with Swedish preschool staff, many might continue to hold superficial views of the links between human inputs and sustainability concerns.

Drawing on our critical analysis above, our extensive backgrounds in ECEfS and our ongoing research engagement, we now outline some of the key theoretical perspectives underpinning ECEfS with a specific focus on theories informing both research and practice. More specifically, we draw on social constructionism, systems theory, critical theory and a rights focus, all of which are strongly linked to early childhood philosophies and policy directives in Australia (Commonwealth DEEWR 2009). In our combined research experience over the last decade (Davis and Elliott 2014), these perspectives offer a depth of possibilities for further progressing ECEfS. However, we do readily acknowledge that this is an emergent field and other theoretical insights will be forthcoming as the research agenda in ECEfS builds.

12.2.1 Sharing Understandings Through Social Constructionism

The social constructionist epistemology invites the researcher to construct "the knowable" in a particular sociocultural and historical context. Notably, the dire consequences of global climate change now sharply define the historical context. Michael Crotty (1998) identifies social constructionism as inclusive, where "the social world and the natural world are not to be seen... as separate distinct worlds existing side by side" (p. 57). This epistemology invites collaborative, socially situated and mediated research approaches where shared understandings can emerge (Elliott 2012). Further, action research methodologies offer valid and productive approaches to research within a social constructionist epistemology and have been well utilised in ECEfS research. The iterative process of social co-construction is central to understandings about sustainability, to shifting ethics about sustainability and to exploring what sustainability may mean for both researchers and educators in early childhood settings. Such an epistemology also underscores current early childhood education philosophy and practice more broadly.

As previously noted, early childhood education philosophy and practice has long been dominated by developmental paradigms until the theorisations of Lev Vygotsky (1978), Urie Bronfenbrenner (1979) and others envisioned children as players in their sociocultural contexts, both impacting on – and being impacted by – their unique and varied sociocultural settings and relationships. Children's development and construction of knowledges about the world are viewed as interwoven within their sociocultural contexts – and in Australia this is core to the EYLF (Commonwealth DEEWR 2009). Constructionism demands the participation of others, shared knowledges and knowledge co-construction with scaffolding by more knowledgeable or skilled peers/adults. This theoretical framing creates an opportunity for working with others to co-construct shared understandings about sustainability and to hypothesise new ways forward in how we (humans) live on the Earth that acknowledges both human sociocultural contexts and the ever-present constraints of our physical global context.

12.2.2 Systems Theory Focussing on Relationships

Systems theory encapsulates a way of thinking that has been recognised over centuries in various cultures, including indigenous ones (Suzuki 2010). As a research methodology, contemporary systems theorists (e.g. Capra 2002) have forged links between the investigation of biological systems and social systems that now inform social science studies, highlighting the relevance of this methodology to sustainability research. As Richard Norgaard (1984) encapsulates, "[humans] modify the ecosystem, while the ecosystem's responses provide cause for individual action and social organisation" (p. 528). The now pressing cause within the co-evolving systems is global sustainability due to excessive human system impacts on the Earth's systems (Commonwealth DCCEE 2011).

Capra (2002) states that systems theory prioritises relationships and that multiple complex and dynamic interdependencies are recognised. Responsiveness is key here and equally pertinent, whether engaging with sustainability or young children and their families, where every element and relationship within the social and biological systems impact on every other one. Both relationships and responsiveness feature in the EYLF to guide educators' engagement with children and families (Commonwealth DEEWR 2009). Also, Bronfenbrenner's (1979) "Ecological Model of Child Development" is often referred to in Australian early education contexts and draws on systems theory to identify the multiple socially constructed systems that impact on children and that, in turn, impact on children as social agents and participants, thus, aligning with the new sociology of childhood (James and Prout 1990). In early childhood education research, a number of studies have documented the potential of systems theory to explore ECEfS (Ferreira and Davis 2010). Also, Young (2009) suggests systems theory offers a non-linear approach to engaging with the big picture of ECEfS, a representation often missing in the pragmatic thinking of educators.

12.2.3 Critical Theory for Making a Difference

Critical theory builds on various discourses and theoretical insights that endeavour to find ways of understanding power and oppression and their impacts on humanity, combined with ways to bring about social change (Freire 1972). Critical theorists question educational knowledge and practices that serve to reproduce the status quo rather than to question and explore issues of power, injustice, marginalisation and inequality (Sung 2007). Also, they view educational facilities such as schools as inextricably linked to issues of power often expressed through, for example, gender, race and ethnicity differentials, and where learners are shaped by standardised, mainstream curriculum and processes.

In other words, critical theorists are suspicious of the constructions that educational settings pass on. A socially critical curriculum, on the other hand, has an emancipatory aim characterised by critical thinking and active and engaged learning designed to "make a difference" here and now. This is in contrast to learning that is seen as preparation for work and later participation in society in adulthood. With respect to EE/EfS, education informed by critical theory not only sees the marginalisation of people as a concern, but also of non-human species, environments and landscapes as a basis for critique and transformation. Within early childhood education contexts, critical theoretical perspectives can influence the curriculum by seeking to deliberately include environmental issues and topics in the everyday learning of young children and involving them in decision-making and action-taking solutions to local problems within their centres, homes and local communities.

12.2.4 A Rights Focus Beyond the Individual

The "United Nations Convention on the Rights of the Child" (UNCRC) (UNICEF 1989) has been both foundational and aspirational since its proclamation. It is a legally binding instrument that incorporates a wide range of human rights for children – civil, cultural, economic, political and social. The UNCRC (UNICEF 1989) recognises children as holders of survival, development, participation and protection rights. Specifically, thinking about participation as a right generates images of children – even the very young – as competent constructors of meanings about the world and their place in it (McNaughton, Hughes and Smith 2008) and aligns with sociology of childhood perspectives mentioned earlier. Children are perceived as social actors with capabilities to participate in matters that affect them, and this lends support to the action-oriented approaches of EfS.

Nevertheless, we argue that even this revisioning of children's rights is not enough to deal with issues of sustainability and that a broader rights framework is called for. Julie Davis (2014) argues children's rights in the age of (un)sustainability need to be extended to include collective rights, intergenerational rights and rights that are beyond human – i.e. bio- or ecocentric rights that assign rights to the Earth's nonliving systems and processes and non-human beings. In the context of early childhood education, such a broadening of rights offers a new rationale for focussing on early education that connects children with the natural world, promotes their understandings of the realities of human–environment interdependencies and supports them as active citizens who can make a difference from an early age.

12.3 Policy Directions Informing Change in Australia

Two Australian policy directives of particular importance for sustainability are the EYLF (Commonwealth DEEWR 2009) and the NQS (ACECQA 2013). We examine both documents below and offer analysis in relation to their impacts and influence on the field's uptake of ECEfS.

12.3.1 Belonging, Being and Becoming

The development of the EYLF (Commonwealth DEEWR 2009) was undertaken by a team of early childhood academics, supported by extensive consultation with practitioners and academics. This consultative process created a unique time and space within the field for discussion about the values, philosophies and theories informing early childhood education practice nationally (Sumsion et al. 2009). While sustainability is not explicitly a core value in the EYLF, there is considerable potential for aligning aspects of the EYLF with the principles of the "Living Sustainably: National Action Plan" (Commonwealth DEWHA 2009), as outlined by Elliott (2014). For example, the EYLF strongly supports naturalised outdoor play settings, "playspaces in natural environments include plants, trees, edible gardens, sand, rocks, mud, water and other elements from nature" (Commonwealth DEEWR 2009, p. 16). However, we still question whether this is sufficient given the sustainability–nature nexus previously elaborated.

The five key children's learning outcomes integral to the framework offer broad descriptions, and while all outcomes have potential to provide opportunities for embedding sustainability (Elliott 2014), many practitioners associate sustainability specifically with outcome 2 (Commonwealth DEEWR 2009, pp. 3–4). The full set of outcomes comprise (1) children have a strong sense of identity; (2) children are connected with and contribute to their world; (3) children have a strong sense of well-being; (4) children are confident and involved learners; and (5) children are effective communicators.

Nevertheless, although the EYLF (Commonwealth DEEWR 2009) is generally well accepted across early childhood education, it has also been subject to critique. While Jennifer Sumsion and Sandie Wong (2011) acknowledge that the key themes of the title, "Belonging, Being and Becoming", are flexible and offer opportunities for interrogation and even possibly radical reinterpretation, Elliott (2014), in particular, has questioned the dominance of sociocultural theory in current interpretations of these themes. As such, humans and human relationships are prioritised over relationships with other species and the Earth, continuing an anthropocentric view of the world. She argues that a biocentric/ecocentric lens is required to reinterpret "Belonging, Being and Becoming" in order for practitioners to fully embrace sustainability. Further, Davis and Arlemalm-Hagser (2014) in their critique of the EYLF and the Swedish preschool curriculum found that both documents really only offer vague and ambiguous definitions and directions with regard to sustainability concepts, values and practices.

12.3.2 National Quality Standard

The NQS, administered by the Australian Children's Education and Care Quality Authority (ACECQA 2013), was launched in 2012 as a quality rating and assessment scheme applicable to all early childhood services in Australia. For the first time and after considerable debate, it aligned existing state-based regulations and national quality areas into one comprehensive document to guide early childhood service quality improvement. The explicit inclusion of sustainability, stated under Quality Area 3 (QA3), was welcomed by early childhood sustainability advocates as the first significant step towards systemic change within early childhood education and was seen as further evidence that ECEfS was "moving forward from the margins", as described below (ACECQA 2013, pp. 99–103):

> Quality Area 3: Physical environment
>
> Standard 3.3: The service takes an active role in caring for its environment and contributes to a sustainable future.
>
> Element 3.3.1: Sustainable practices are embedded in service operations.
>
> Element 3.3.2: Children are supported to become environmentally responsible and show respect for the environment

A spate of practitioner publications (Sneddon and Pettit 2016; Young and Elliott 2014) and professional learning sessions supported the implementation of QA3.3. In particular, one of Australia's state-based ECEfS professional networks, the New South Wales Early Childhood Environmental Education Network (NSW ECEEN), secured government funding to develop a guide to embedding sustainability into all quality areas of the NQS, called "Ecosmart for early childhood: A sustainability filter for quality improvement plans" (NSW/ECEEN and NSW/OEH 2012). The guide challenges and extends practitioners' perceptions that sustainability is only relevant to QA3.3.

Amidst these positive steps towards embedding EfS into early education practice in Australia, it is evident that there remain significant challenges within the field regarding the implementation of QA3.3. Anecdotally, practitioners and NQS assessors – those experienced education professionals whose task it is to determine whether or not a service does actually meet the regulated quality standards – are asking "What is sustainability?" and "What might a service meeting or exceeding the QA3.3 Standard look like?" A somewhat colloquial interpretation is that establishing a compost bin or vegetable garden will suffice to meet the standard. This, perhaps, reflects the positioning of sustainability within QA3: physical environment that, whether intended or not, situates sustainability as a physical environment issue, rather than being interpreted as a concept that requires significant and transformative social change across multiple dimensions. As ACECQA regularly reviews the assessments and ratings, it has identified QA3.3 to be problematic for many services, particularly in some states and territories (ACECQA 2015). From the fairly limited data publicly available, we speculate that it is in the states and territories where there are long-standing professional early childhood sustainability networks dating back to as early as 1992 that QA3.3 is rated more highly. We have acknowledged elsewhere that these networks are unique and influential in the Australian ECEfS landscape (Davis and Elliott 2014).

The challenges that QA3.3 present to the early childhood field have not gone unnoticed. Most recently, a government review of the NQS (Productivity Commission 2014) proposed that sustainability be fully removed from the NQS on the grounds that it is proving too difficult for educators to implement and should not be a part of government-legislated quality criteria. Hence from late 2017, the sustainability standard will be replaced with an environmental responsibility element; and, we now question if being sustainable and educating for sustainability in early childhood services may revert to being optional and marginal. From our perspective, this is extremely short-sighted political thinking pedalled by the current government which has overly responded to right-wing free enterprise businesses such as for-profit, childcare centres. The pervasiveness of such neo-liberal thinking is evident in many spheres of education and includes a proposal to remove sustainability as a cross-curriculum thread from the Australian school curriculum (Australian Curriculum and Reporting Authority 2014). The clear intent is to weaken the already rather weak references to sustainability across the entire education spectrum. It remains to be seen whether the gradual, but solid, uptake by practitioners within the early childhood field, and continued advocacy by ECEfS networks and professional associations for inclusion of EfS, is strong enough to withstand such a watering down of policy. It is important that EfS advocates remain alert to possible future attacks on sustainability in the national curriculum.

12.4 Current Research: Emerging Tensions and Challenges

To summarise, despite the somewhat positive overall shifts in EfS practice and policy, a recent government-funded consultative study (Elliott et al. 2016) has identified emerging tensions and challenges for the implementation of ECEfS in Australian early education services. Here, we highlight two specific foci: a lack of sustainability knowledge and related professional learning issues amongst practitioners and the importance of governance, leadership and systemic change across the broader early childhood education field. As long-term advocates, we also recognise that these priorities have been on the Australian ECEfS agenda for many years (NSW EPA 2003).

12.4.1 Demystifying Sustainability: Practitioner Knowledge and Professional Learning

As identified in relation to the NQS (ACECQA 2013), sustainability remains a topic requiring demystification for many practitioners. Recent studies have demonstrated that this is a challenging construct and invariably sustainability is described by early childhood practitioners as nature-based activities or tangible sustainable practices such as installing water tanks and worm farms (Hill et al. 2014). A "tick-the-sustainability-box" mentality pervades, most probably reinforced by the NQS rating process (ACECQA 2013) when what is required instead is deep transformative change such that sustainability becomes the "norm" or simply "the way we do things". The following quote from an early childhood practitioner phone interview aptly describes this issue:

> We have focused very much on sustainability within the environment and not looking at sustainability as a whole... we automatically link sustainability with recycling and that kind of thing and yet... we need to look at it more holistically... and on a global scale. That's where there a lot of gaps and misconceptions about what sustainability actually entails. It's more than just having your compost bin... there are gaps in the knowledge that we have. (Elliott et al. 2016, p. 81)

These findings are not unique to early childhood education, however, and correlate with a lack of awareness and comprehension also cited as barriers to EfS uptake in the Australian school sector (AESA 2014).

Appropriate and widespread professional learning in ECEfS at all levels is essential to addressing this lack of sustainability knowledge. Although some professional organisations such as Gowrie Australia and KU Children's Services have implemented in-service training initiatives, currently, there is no co-ordinated national approach to professional learning in sustainability with little or no way of knowing whether what is offered simply adds to the superficiality of strategies such as nature play or recycling activities. Further, Elliott and Nadine McCrea (2015) have identified that ECEfS knowledge alone may be insufficient and many educators also require guidance to translate ECEfS knowledge into authentic pedagogical practice with children. The links between ECEfS knowledge and pedagogy and the proposed "pedagogical advantage" (Elliott and Davis 2009) of early childhood are not self-evident to many practitioners. Professional learning must also be offered in a range of formats from online webinars to fact sheets, local learning hubs and face-to-face workshops. No one format suits all, and the constraints of working hours and travel distances in vast regional areas must be considered (Dyment et al. 2014). Also, it is well recognised that there is a range of early childhood qualifications, pedagogical experience, sustainability knowledge and values amongst practitioners; hence, multiple starting points for professional learning must be accommodated (Elliott and McCrea 2015).

Moreover, in teacher education, as yet, there is no picture of what is offered within the many universities that provide early childhood teacher education programmes; one presumes the image remains one of "patches of green". While several recently developed early childhood teacher education courses are known to include sustainability units, this is not the norm, and there is no legislative requirement to do so (Elliott et al. 2016). By contrast, the technical and further education vocational sector that offers diploma and certificate-level children's services studies has had training packages focussed on sustainability nationally since 2012 (TAFE NSW Training and Education Support Industry Skills Unit 2010). Already some services are noting the impacts of this training in building a critical mass for service culture change, but there are many practitioners who were trained pre-2012 and are yet to professionally engage with sustainability (Elliott and McCrea 2015).

12.4.2 Governance, Leadership and Systemic Change

The topics of governance, leadership and systemic organisational change are key to progressing and strengthening ECEfS uptake (Elliott et al. 2016). However, it was evident in this study that the professional early childhood management bodies canvassed were only just beginning to consider sustainability as relevant to organisational mission statements, strategic plans, policies or member service philosophies. Systemic change was seen as feasible only when these aspects of governance aligned with sustainability for all early childhood member services to implement as relevant in their settings. Study participant practitioners indicated they were seeking leadership from government, their professional management organisations and service co-ordinators. However, they stressed ownership of change, rather than imposed change towards sustainability was critical. As the early childhood field has experienced significant government-instigated policy change at many levels over the last decade (COAG 2008), change management itself and viewing sustainability as a dynamic process of continuing and deepening change were considered a challenge for services. Yet, incremental change over time is core to embedded and ongoing ECEfS values and pedagogy.

Likewise, leadership is a vexed and much debated topic in the highly gendered early childhood profession and is often perceived in practice as more about day-to-day management. One way to address this issue was the creation of the role of "pedagogical leader" across the early childhood field at the same time as the NQS was legislated. A current requirement for early childhood services in Australia is to have at least one educational/ pedagogical leader per location. This person "leads the development of the curriculum and ensures the establishment of clear goals and expectations for teaching and learning" (ACECQA undated online). With the emergence of this position, interest in extending knowledge about leadership more generally has arisen, with potential for building in EfS if educational leaders are appropriately knowledgeable about sustainability. Further, as Christine Woodrow and Gillian Busch (2008, p. 90) argue, "activist professional leaders [capable of] embracing conflict and difference as necessary for change and growth in relationships" are required if ECEfS is to have robust and vocal leaders to progress the field into the future.

12.5 Moving Forward

This chapter highlights recent significant progress in ECEfS in Australia that may inform other countries seeking both policy and pragmatic directions for promoting EfS in early childhood education contexts. In recent times, there has been a positive surge in uptake, with ECEfS being recognised in the UN DESD end of decade report (UNESCO 2014) as having made significant strides, although realistically the baseline was very low. We particularly highlight that meanings of sustainability and interpretations in practice vary markedly across majority and minority global worlds – there are inequities and injustices to be addressed as integral to engaging in sustainability globally. Nevertheless, we argue that, in many ways, the stage is ready for children to "do more" than nature activities; they need to be supported in their early education contexts to become environmentally knowledgeable and skilful. Yet, children's environmental agency is still a major work in progress (Elliott et al. 2016). More so, how children learn about sustainability is recognised as a significant research gap (Hedeflak et al. 2014), and we believe that a critically reflective pedagogy is fundamental to embracing sustainability with children. ECEfS remains a nascent field that needs nurturing if it is to continue expanding and "moving forward from the margins". Engagement with ECEfS, both internationally and in Australia, requires ongoing and multiple research studies and collaborations, and both politically-driven and service-driven systemic approaches, so that early education can fully contribute to creating the globally inclusive and sustainable futures essential for all generations.

Questions

Discussion questions that could help instructors and students to engage in meaningful conversation about the ideas presented in this chapter.

1. How is education for sustainability evident, if at all, in policy and practice in universities, schools and early childhood education services within your country?
2. How do you reconcile the UNESCO conception of sustainability as multidimensional with a focus on nature-based activities that pervade early education?
3. When considering the four theoretical perspectives discussed in the chapter, how do/might these guide implementation of education for sustainability in your contexts?
4. What organizations or government agencies could you partner with to advocate for enhancing early childhood education for sustainability provision?

References

Ärlemalm-Hagsér, E., & Sandberg, A. (2011). Sustainable development in early childhood education: In–service students' comprehension of SD. *Environmental Education Research Journal, 17*(2), 187–200.

Australian Children's Education & Care Quality Authority (ACECQA). (2013). *Guide to the national standard.* Retrieved from http://files.acecqa.gov.au/files/National–Quality–Framework–Resources–Kit/NQF03–Guide–to–NQS–130902.pdf

Australian Children's Education & Care Quality Authority (ACECQA). (2015). NQF snapshot Q4 2014. Retrieved from http://files.acecqa.gov.au/files/Reports/2015/Feb_2015_ACECQA_Snapshot_Q4_FINAL_WEB.pdf

Australian Children's Education & Care Quality Authority (ACECQA). (n.d.). *The role of the educational leader.* Retrieved 20 Febraury 2017 from, http://files.acecqa.gov.au/files/Information%20sheets/Portal%20–%20The%20role%20of%20the%20educational%20leader%20–%20information%20sheet%20PDF%20version.pdf

Australian Curriculum and Reporting Authority (ACARA). (2014). Australian curriculum V5.2. Retrieved 25 April 2015 from www.acara.edu.au.
Australian Education for Sustainability Alliance (AESA). (2014). *Education for sustainability and the Australian curriculum project: Final report for research phases 1 to 3*. Melbourne: AESA.
Bronfenbrenner, U. (1979). *The ecology of human development*. Cambridge, MA: Harvard University Press.
Brooker, L., & Edwards, S. (Eds.). (2010). *Engaging play*. Maidenhead: Open University Press.
Calman, L., & Tarr–Whelan, L. (2005). *Early childhood education for all: A wise investment*. New York: Momentum.
Capra, F. (2002). *The hidden connections*. London: Flamingo.
Commonwealth of Australia Department of Climate Change and Energy Efficiency (DCCEE). (2011). *Australian climate commission report: The critical decade*. Canberra: Commonwealth Department of Climate Change and Energy Efficiency.
Commonwealth of Australia Department of Education, Employment & Workplace Relations (DEEWR). (2009). *Belonging, being and becoming: The early years learning framework for Australia*. Canberra: Department of Education, Employment & Workplace Relations.
Commonwealth of Australia Department of Environment Water Heritage & the Arts (DEWHA). (2009). *Living sustainably: National action plan*. Canberra: Commonwealth Department of Environment Water Heritage & the Arts.
Council of Australian Governments (COAG). (2008). *A national quality framework for early childhood education and care: A discussion paper*. Canberra: COAG.
Crotty, M. (1998). *The foundation of social research: Meaning and perspective in the research process*. Allen & Unwin: Crow's Nest.
Davis, J. (1999). Early childhood environmental education—Mainstream not marginal. *OzEEnews, 76*, 2.
Davis, J. (2007). Climate change and its impact on young children. *Early childhood Australia*. Retrieved from www.eca.org.au
Davis, J. (2009). Revealing the research 'hole' of early childhood education for sustainability: A preliminary survey of the literature. *Environmental Education Research, 15*(2), 227–241.
Davis, J. (2014). Examining early childhood education through the lens of education for sustainability: Revisioning rights. In J. Davis & S. Elliott (Eds.), *Research in early childhood education for sustainability: International perspectives and provocations* (pp. 21–37). London: Routledge.
Davis, J. (Ed.). (2015). *Young children and the environment: Early education for sustainability* (2nd ed.). Port Melbourne: Cambridge University Press.
Davis, J., & Arlemalm–Hagser, E. (2014). Examining the rhetoric: A comparison of how sustainability and young children's participation and agency are framed in Australian and Swedish early childhood education curricula. *Contemporary Issues in Early Childhood, 15*(3), 231–244.
Davis, J., & Elliott, S. (2014). *Research in early childhood education for sustainability: International perspectives and provocations*. London: Routledge.
Dickinson, E. (2013). The misdiagnosis: Rethinking "nature deficit disorder". *Environmental Communication, 7*(3), 315–355. https://doi.org/10.1080/17524032.2013.802704.
Dyment, J., Davis, J., Nailon, D., Emery, S., Getenet, S., McCrea, N., & Hill, A. (2014). The impact of professional development on early childhood educators' confidence, understanding and knowledge of education for sustainability. *Environmental Education Research, 20*(5), 660–679.
Elliott, S. (2012). *Sustainable outdoor playspaces in early childhood settings: Investigating perceptions, facilitating change and generating theory*. Unpublished doctoral thesis, University of New England.
Elliott, S. (2014). *Sustainability and the early years learning framework*. Mt Victoria: Pademelon Press.
Elliott, S., & Davis, J. (2009). Exploring the resistance: An Australian perspective on educating for sustainability in early childhood. *International Journal of Early Childhood, 41*(2), 65–77.

Elliott, S., & McCrea, N. (2015). Gaps and challenges informing professional learning about early childhood education for sustainability. *The Social Educator, 33*(3), 17–28.

Elliott, S., & Young, T. (2015). Nature by default in early childhood education for sustainability. *Australian Journal of Environmental Education, 32*(1), 57–64. https://doi.org/10.1017/aee.2015.44.

Elliott, S., McCrea, N., Newsome, L., & Gaul, J. (2016). Examining environmental education in NSW early childhood education services: A literature review with findings from the field. Sydney, Australia: NSW Office of Environment & Heritage Environmental Trust. (This study and report was funded by the NSW Office of Environment & Heritage Environmental Trust). Retrieved from http://www.environment.nsw.gov.au/grants/dissemination.htm

Engdahl, I., & Arlemalm–Hagser, E. (2014). Education for sustainability in Swedish preschools: Stepping forward or out–of–step? In J. Davis & S. Elliott (Eds.), *Research in early childhood education for sustainability: International perspectives and provocations* (pp. 208–224). London: Routledge.

Feriver, S., Teksöz, G., Olgan, R., & Reid, A. (2015). Training early childhood teachers for sustainability: Towards a 'learning experience of a different kind'. *Environmental Education Research, 22*(5), 717–746. https://doi.org/10.1080/13504622.2015.1027883.

Ferreria, J., & Davis, J. (2010). Creating deep and broad change through research and systems approaches in early childhood education for sustainability. In J. Davis (Ed.), *Young children and the environment: Early education for sustainability* (pp. 273–289). Melbourne: Cambridge Press.

Freire, P. (1972). *Pedagogy for the oppressed.* Harmondsworth: Penguin.

Gough, A. (1997). *Education and the environment.* Melbourne: Australian Council for Educational Research.

Hedefalk, M., Almqvist, J., & Östman, L. (2014). Education for sustainable development in early childhood education: A review of the research literature. *Environmental Education Research, 21*(7), 1–16. https://doi.org/10.1080/13504622.2014.971716.

Hill, A., McCrea, N., Emery, S., Nailon, D., Davis, J., Dyment, J., & Getenet, S. (2014). Exploring how adults who work with young children conceptualise sustainability and describe their practice initiatives. *Australasian Journal of Early Childhood, 39*(3), 14–22.

James, A., & Prout, A. (1990). *Constructing and reconstructing childhood.* London: Falmer Press.

Knight, S. (2013). *International perspectives on forest school: Natural places to play and learn.* London: Sage Publications.

Louv, R. (2005). *The last child in the woods: Saving our children from nature deficit disorder.* Chapel Hill: Algonquin Books.

MacNaughton, G., Hughes, P., & Smith, K. (Eds.). (2008). *Young children as active citizens: Principles, policies and pedagogies.* Newcastle–on–Tyne: Cambridge Scholars Publishing.

McNichol, H., Davis, J. M., O'Brien, K., & R. (2011). An ecological footprint for an early learning centre: Identifying opportunities for early childhood sustainability education through interdisciplinary research. *Environmental Education Research, 17*(5), 689–704.

Norgaard, R. B. (1984). Co–evolutionary agricultural development. *Economic Development and Cultural Change, 32*(3), 525–546.

NSW ECEEN & NSW OEH. (2012). *Ecosmart for early childhood: A sustainability filter for quality improvement plans.* Sydney: NSW ECEEN & NSW EOH.

NSW Environment Protection Authority (NSW EPA). (2003). *Patches of green: A review of early childhood environmental education.* Sydney: NSW EPA.

Organisation for Economic Co–operation and Development (OECD). (2006). *Starting strong 11: Early childhood education and care.* Paris: OECD.

Organisation for Economic Co–operation and Development (OECD). (2012). *Starting strong III: A quality toolbox for early childhood education and care.* Paris: OECD.

Productivity Commission. (2014). *Childcare and early childhood learning. Inquiry Report No. 73.* Canberra: Productivity Commission.

Shonkoff, J. P., & Richter, L. (2013). The powerful reach of early childhood development: A science based foundation for sound investment. In P. R. Britto, P. L. Engle, & C. S. Super (Eds.), *Handbook of early childhood development research and its impact on global policy* (pp. 1–23). New York: Oxford University Press.

Sneddon, S., & Pettit, A. (2016). *Sustainability in action in early childhood settings*. Blairgowrie: Teaching Solutions.

Sumsion, J., & Wong, S. (2011). Interrogating 'belonging' in belonging, being and becoming: The early years learning framework for Australia. *Contemporary Issues in Early Childhood, 12*(1), 28–45.

Sumsion, J., Barnes, S., Cheeseman, S., Harrison, L., Kennedy, A., & Stonehouse, A. (2009). Insider perspectives on developing belonging, being and becoming: The early years learning framework for Australia. *Australasian Journal of Early Childhood, 34*(4), 4–13.

Sung, K. (2007). Globalising critical pedagogy: A case study of English language teaching in Korea. In P. McLaren & J. Kincheloe (Eds.), *Critical pedagogy: Where are we now?* New York: Peter Lang.

Suzuki, D. (2010). *The legacy: An elder's vision for a sustainable future*. Crow's Nest: Allen & Unwin.

Taylor, A. (2013). *Reconfiguring the natures of childhood*. London: Routledge.

Technical & Further Education (TAFE) New South Wales Training & Education Support Industry Skills Unit. (2010). *Small footprints: CHCPR515A develop and implement a program to support sustainable practice*. Meadowbank: TAFE NSW Training & Education Support Industry Skills Unit.

UNESCO. (2005). *Decade of education for sustainable development 2005–2014: Draft international implementation scheme*. Paris: UNESCO.

UNESCO. (2008a). *Early childhood and its contribution to a sustainable society*. Paris: UNESCO.

UNESCO. (2008b). *The Gothenberg recommendations on education for sustainable development*. Paris: UNESCO.

UNESCO. (2010). *Four dimensions of sustainable development*. Retrieved from http://www.unesco.org/education/tlsf/mods/theme_a/popups/mod04t01s03.html

UNESCO. (2014). *Shaping the future we want: UN decade of education for sustainable development (2005–2014). Final Report*. Paris: UNESCO.

UNESCO UNEP. (1977). *Intergovernmental conference on environmental education: Tbilisi (USSR)*, 14–26 October 1997. Final Report. Paris: UNESCO.

UNICEF. (1989). *United Nations Convention of the Rights of the Child (UNCRC)*. Retrieved from http://www.ohchr.org/en/professionalinterest/pages/crc.aspx

Vygotsky, L. (1978). The role of play in development. In *Mind in society* (M. Cole, Trans., pp. 92–104). Cambridge, MA: Harvard University Press.

Wilson, R. (1993). *Fostering a sense of wonder during the early childhood years*. Columbus: Greyden Press.

Woodrow, C., & Busch, G. (2008). Repositioning early childhood leadership as action and activism. *European Early Childhood Research Journal, 16*(1), 83–93.

Young, T. (2009). *"The crack where the light gets in": An examination of sustainability education in early childhood education*. Unpublished master's thesis, Monash University.

Young, T., & Elliott, S. (2014). *Ways of thinking, acting and relating about sustainability*. Deakin West: Early Childhood Australia.

Chapter 13
Environmental Education in China: A Case Study of Four Elementary and Secondary Schools

Qin Chengqiang, Xiong Ying, Feng Yan, and Li Tian

Abstract As a country with a fast-growing economy and a large population, China has shown an increasing concern over environmental issues since the 1972 United Nations Conference on the Human Environment. As a result, environmental education (EE) in Chinese elementary and secondary schools has achieved remarkable progress. However, the current effects of environmental damage seem to outweigh the positive effects of decades-long school-based EE initiatives in the country. Based on a survey conducted with four elementary and secondary schools in Nanning – known as the Green City – the present chapter reveals important barriers to the improvement of EE in the Chinese educational system. We conclude our discussion by making several recommendations to address those obstacles identified by participants in our study which are common to other countries as well.

13.1 What Are the Problems?

Since the carrying out of the Opening and Reform Policy in 1978, the Chinese government has loosened its tight control over certain economic activities, including international trade. As a result, China has maintained a high average gross domestic product rate and is recognized as the second-largest economy of the world today (Han 2015). In addition, according to the National Bureau of Statistics (2012), the country's GDP per capita is ranked 100th globally. At the same time, approximately 122 million people live in poverty (National Development and Reform Commission 2012), and the rapid process of industrialization and urbanization has caused increasing depletion of natural resources, thus putting huge pressure on the environment. As a result, any current prospects of sustainability in China are

Q. Chengqiang (✉) • X. Ying • F. Yan • L. Tian
Guangxi University,
100 Daxue Road, X ixiangtang District, Nanning, Guangxi 530004, China
e-mail: cqqin@gxu.edu.cn; xy2002@gxu.edu.cn; fengyanscic@qq.com; litianpsyc@qq.com

G. Reis, J. Scott (eds.), *International Perspectives on the Theory and Practice of Environmental Education: A Reader*, Environmental Discourses in Science Education 3, https://doi.org/10.1007/978-3-319-67732-3_13

daunting. Nevertheless, since the 1990s, education has been recognized as an indispensable and essential approach to raising public awareness toward attaining sustainable development (Han 2015). Accordingly, environmental education (EE) in elementary and secondary schools has received increasing attention in the Chinese educational system, specifically after the release of UN's Agenda 21 (Tang 1994).

The importance of EE in the elementary and secondary education is recognized by the Chinese society and educators through the variety of EE textbooks published and the many training opportunities offered to elementary and secondary teachers. However, these initiatives were not enough to stop the environmental crisis in China. During the last decade, ecosystems have continually degenerated, soil erosion has become serious, the number of endangered species has increased, and many natural wetlands have disappeared. What has happened? Why hasn't school-based EE been able to help alleviate the environmental damage being caused in the country? In order to find out, a survey was conducted with four elementary and secondary schools in Nanning, considered to be the "Green City" of China. It is the capital city of Guangxi Zhuang Autonomous Region, thus being the provincial center of politics, economy, science, technology, education, culture, and health (Nanning Government website 2015). The city also plays an important role in the economic development of Southwest China and is gradually becoming a commercial and communication center for Southeast Asia (Nanning Government website 2015).

The purpose of this chapter is to explore the significance of the results of this EE survey and how they can help inform EE-led initiatives in Chinese schools and internationally.

13.2 A Brief Review of History: EE Policies in China

This section provides context for the various policy reforms implemented in the country over the last four decades and which have influenced the presence of EE in Chinese schools to this day. The roots of modern EE in China began to take hold in the early 1970s and have gone through three periods of development, explained as follows:

13.2.1 The Start-Up Period: 1972–1988

In 1972, the Chinese government sent a delegation to attend the United Nations Conference on the Human Environment held in Sweden. During the following year, China's first National Environmental Protection Conference, held in Beijing, highlighted that universities and colleges should offer environmental protection (EP) programs (Li 2004). Following that conference, Tsinghua University and Peking University set up the first environmental engineering undergraduate and postgraduate degree programs in China, respectively.

In 1978, the Environmental Protection Leading Group of the State Council proposed the integration of EP into the curriculum for elementary and secondary schools. In the following year, the promulgation of the Environmental Protection Law of the People's Republic of China highlighted the importance of disseminating the EP more broadly. The People's Education Press began to incorporate EP into natural science, geography, and chemistry textbooks for elementary and secondary school grades. Promoted by the Chinese Society for Environmental Sciences (CSES), EP pilot projects were carried out in selected schools. Two years later, the CSES expanded their projects to more provinces and cities and started offering EP training workshops for teachers. In the same year, EP was acknowledged as one of China's basic national policies, which contributed to making EE an integral part of its development until 1988.

13.2.2 The Steadily Developing Period: 1989–1995

In 1989, the symposium jointly held by the State Environmental Protection Administration (SEPA) and CSES promoted the systematization, standardization, and regularization of EE in basic education (Li 2003). Participants at this conference recognized the importance of nongovernmental financial support for school-based EE, more training opportunities for teachers, and the production of higher-quality textbooks. In 1992, the State Education Commission (SEC) explicitly required schools to incorporate EE into their curriculum through the policy document "Education is the Foundation for EP," which was put forward at the 1st National Environmental Education Working Conference.

In 1994, 2 years after the first United Nations Conference on Environment and Development in Rio de Janeiro, the Chinese government issued "China's Agenda 21: White Paper on China's Population, Environment and Development in the 21st Century." It claimed that sustainable development (SD) was a top priority in China's development and reform agenda (Han 2015). This policy document was also viewed as a fundamental in guiding China's social and economic development (State Council 1994).

13.2.3 The Flourishing Period: 1996 – Present

In 1996, China initiated the National Environmental Publicity and Education Program (1996–2010), which served to connect EE with the quality education agenda (Han 2015) and laid the foundations for a nationwide EE program. The state-supported Green School initiative was launched in the same period to encourage schools at all levels throughout China to get involved in EE. In order to promote the development of Green Schools, the Ministry of Education (MOE) awarded a national commendation to Green Schools in 2000 and presented this recognition to

National Green Schools every 2 years. In 1997, the MOE established the nationwide Environmental Educators' Initiative (EEI) in collaboration with the World Wildlife Fund (WWF) and British Petroleum.

The year 2003 was exciting for EE in China due to the issuance of the first guiding policy on EE by the MOE, "Guidelines for Implementing Environmental Education in Elementary and Secondary Schools," which drew from the experience of EEI. The Guidelines marked the increasing importance of education for environment and sustainability in Chinese elementary and secondary education system and have greatly promoted the integration of EE into specific school programs. As a result, the MOE was thus awarded the Gifts to the Earth by WWF in 2003. In the same year, the Chinese government put forward the Scientific Outlook on Development (SOD), recognized as an important theoretical guidance for developing EE in the nation. It states:

> To conduct sustainable development, China shall foster a harmony between man and nature, tackle problems inherent in economic construction, population growth, resource utilization and environmental protection, and push society onto a path toward civilized development featuring growing production, an affluent lifestyle and a sound ecosystem. (China Daily 2007)

In 2011, the National Environmental Publicity and Education Program (2011–2015) encouraged the establishment of social practice bases in collaboration of schools with botanic gardens, science parks, museums, laboratories of academic institutions, nongovernmental EP organizations, and the like. Two years later, 80 national-level bases were approved by the MOE. These have greatly changed the EE landscape, developed students' environmental literacy, fostered their awareness of and concern about ecological civilization and SD, and created new patterns to provide students with opportunities to acquire the knowledge, values, attitudes, commitments, and skills needed to protect and improve the environment.

Over the past four decades, the Chinese government's efforts to make EE an integral part of elementary and secondary education through its national curriculum framework is clear through the development of new strategies for the design and implementation of EE instructional materials for both elementary and secondary education system. Although a nationwide, multilevel, and multiform EE system has been established, it has been influenced by international initiatives that are insensitive to the local contexts where it has been implemented. We therefore conducted a survey with students and teachers to examine the problems of implementing EE in the educational system at two elementary (grades 4–6) and two secondary (grades 7–9) schools in Nanning, also called The Green City because of its abundance of plants. The student questionnaire included three categories for examination (environmental knowledge, environmental awareness, and environmental education), while the teachers' questionnaire contained two categories (environmental awareness and teaching methodologies).

13.3 Background of the Study

Three research questions guided our study: Can current EE teaching methodologies within the elementary and secondary schools improve children's environmental awareness? Has the current EE system within the elementary and secondary schools been evaluated by the government effectively? Are EE programs within the elementary and secondary schools fully supported by the government?

In total, 1200 student questionnaires were distributed: 300 questionnaires to each school. 216 questionnaires were fully completed and returned, representing a response rate of 18%. Of those that returned the survey, 121 participants were elementary students (56.0%), while 95 were secondary students (44.0%). Males comprised 44.9% of our sample.

We distributed 120 teacher questionnaires in these four schools. Forty questionnaires were fully completed and returned, representing a response rate of 33.3%. Nineteen elementary teachers completed and returned their survey (47.5%), while 21 secondary returned the survey (52.5%). Male teacher participants comprised 15% of our sample. The average teaching tenure within the school was 12.8 years. Thirty-three participants (82.5%) indicated they held a bachelor degree, whereas only four participants (10%) possessed a teaching diploma. Three teachers (7.5%) held a master's degree. Based on teachers work experiences and tenure, they are ranked in three different levels (junior professional title, intermediate professional title, and senior professional title) by China's Ministry of Education. Twenty-seven participants (67.5%) reported they had an intermediate professional title. Ten participants (25%) held a junior professional title, and three participants (7.5%) obtained a senior professional title in the survey.

Seven teachers taught grade four (17.5%), the same being true for grade six. Five participants were found to teach each of the following grades: five, seven, and nine (12.5% each). The other 11 respondents were from grade eight, which represented the highest respondent percentage (27.5%). Regarding the teaching subjects, Chinese comprised 32.5%, followed by mathematics (22.5%), English (17.5%), art (7.5%), history (7.5%), physics (5%), and geography, chemistry, and politics (2.5% each). Twenty-eight participants (70%) reported they had not taken any preservice EE training before beginning to teach. Meanwhile, 25 (62.5%) participants indicated they had not received any in-service EE training (Table 13.1).

The two paper-based survey questionnaires were designed by the Sino-Canadian Comparative Environmental Research Unit of Guangxi University. The student questionnaire included three variables: environmental knowledge ($\alpha = 0.63$), environmental awareness ($\alpha = 0.68$), and environmental education ($\alpha = 0.64$). The teacher questionnaire contained two variables: environmental awareness ($\alpha = 0.71$) and teaching methodologies ($\alpha = 0.85$).

Included in the teacher survey, three open-ended questions were provided in order to explore concerns for EE development in the future within the educational system. For example, What are and will be the biggest problems for teaching EE in China? In order to improve EE instruction in elementary and secondary schools, what teaching methodologies should teachers focus on? We also asked participants to provide recommendations for promoting EE in the elementary and secondary schools.

Table 13.1 Number of participants from each school

School code	Students questionnaires distributed	Number of respondents	Response rate	Teachers questionnaires distributed	Number of respondents	Response rate
Elementary 1	300	61	20.3%	30	8	26.7%
Elementary 2	300	60	20.0%	30	11	36.7%
Secondary 3	300	55	18.3%	30	10	33.3%
Secondary 4	300	40	13.3%	30	11	36.7%
Total	1200	216	18.0%	120	40	33.3%

In 1976, the Belgrade Charter declared significant educational efforts were required to improve life and protect the environment. At Tbilisi, the UNESCO endeavored to expand on the efforts made in Belgrade with the development of new EE categories (The Tbilisi Declaration 1978):

- Awareness: to help social groups and individuals acquire an awareness and sensitivity to the total environment and its allied problems
- Knowledge: to help social groups and individuals gain a variety of experience in, and acquire a basic understanding of, the environment and its associated problems
- Attitudes: to help social groups and individuals acquire a set of values and feelings of concern for the environment and the motivation for actively participating in environmental improvement and protection
- Skills: to help social groups and individuals acquire the skills for identifying and solving environmental problems
- Participation: to provide social groups and individuals with an opportunity to be actively involved at all levels in working toward resolution of environmental problems

These categories were used to organize the responses received from students and teachers.

13.3.1 Awareness

A majority of the student respondents (60.7%) reported they were moderately satisfied with their current school environment, while few students were ambivalent with the environment. They also reported to moderately pay attention to media information related to environmental protection. For example, 60% of participants stated that they attempt to save water in their daily lives while 40% indicated they did not pay attention to this issue. The majority (76.7%) claimed that environmental protection is very important and must be protected by everyone right now. However, these the same individuals claimed that they would throw away used paper, bottles, or empty lunch boxes in a public place if they could not find a trash can and while nobody was watching.

In the case of teachers, a majority of them (62.3%) reported that their environmental awareness was at moderate level. They strongly agreed that EE taught in schools is the essential, comprehensive, and effective path to solve environmental issues in China. They also strongly agreed that China's new curriculum reform should improve students' environmental awareness, by integrating EE into the curricula of elementary and secondary school programs.

13.3.2 Knowledge

Most of participants (83.6%) knew the exact date of the World Environment Day and understood that using biodegradable plastic bags could reduce white pollution. Many of participants (52.2%) recognized that people's attitude toward environmental protection is closely related to the level of environmental pollution. Therefore, they pointed out the bicycle was the friendliest transport for our environment. As for teachers, they showed limited knowledge about EE; most of whom (96.4%) indicated that EE is mainly related to natural environment protection.

13.3.3 Attitudes

Results of the student survey reveal that the participants had fewer concerns and low motivation to find out what the main global environmental problems are. While climate change was a main environmental issue that concerned most of the participants, they weren't noticeably concerned about air pollution, water pollution, and/or noise pollution. They demonstrated low motivation to understand issues around waste disposal problems (classification of waste, waste recycling, and reuse), land desertification, droughts and floods, and biodiversity damage to name a few examples.

82.8% of teacher participants disagreed that introducing EE to elementary and secondary schools would negatively affect current Chinese curriculum of education and influence students' high school entrance examination. However, the same participants purported that EE is of little relevance to the courses they were teaching at the time of the study.

13.3.4 Skills

The results also displayed that the participants lack basic skills to identify and solve environmental problems. Most of participants (96.7%) reported that they normally received environmental information via traditional media including newspaper, broadcast, and television. The participants also indicated they generally gained relevant skills through EE activities in their schools, such as environmental knowledge competitions. Many of the respondents (62.6%) reported that classroom teaching, the Internet, or influence from others was not a main path to gain environmental skills.

The vast majority of teachers (94.3%) strongly agreed with the statement: “every elementary and secondary school teacher should understand the principles, contents and methods of environmental education.” However, most of them claimed they knew little about the content of the “Guidelines for the Implementation of Environmental Education in Elementary and Secondary Schools.” They also pointed out that it was necessary to include EE in the teachers’ continuing education courses in the future.

13.3.5 Participation

Seventy-seven percent of participants stated that the most effective way to implement EE is through specific activities (e.g., tree planting). Many participants (91.7%) asserted that the best way to promote EE would be offering specific EE courses. Likewise, teaching environmental knowledge through different courses, watching green videos, having environmental education seminars, or visiting garbage treatment station would be beneficial. However, most only took EE courses, rarely participating in other activities. They also maintained that offering EE courses would not add weight to their studies in school. Currently, most teachers causally teach environmental knowledge to students resulting in less demonstrated interest in EE activities in schools for the respondents.

Many participants (86.5%) also reported that they often presented environmental knowledge in their teachings. However, the participants admit that they rarely provided extra teaching activities related to EE nor did they self-study EE materials or discuss EE issues with other subject teachers. They also never took students to complete specific environmental activities including visiting other environmental organizations.

Similar to the results of student survey, most of teacher participants (96.3%) considered the best way to promote EE in schools would be to offer EE courses and visiting environmental organizations. It was apparent that school administration placed a high premium on EE. At the end of the teacher survey, three open-ended questions were presented to the participants. The most frequent answers to the questions are summarized as follows:

What are and will be the biggest problems for teaching EE in China?

1. Students do not have enough environmental awareness.
2. EE courses and practices are unrelated, because many things are empty talks.
3. There is an urgent need to connect EE and teaching materials.
4. Currently there are huge difficulties in organizing EE activities after school.
5. Teachers have not been trained systematically in the implementation of EE.

In order to improve EE instruction in elementary and secondary schools, what teaching methodologies should teachers focus on?

1. Visiting and studying from environmental organizations.
2. Providing professional development opportunities.
3. Providing environmental education seminars.
4. Participating in EE activities with students after school.

Provide recommendations for promoting EE in the elementary and secondary schools.

1. Offering a regular EE course in elementary and secondary school.
2. Teaching and integrating environmental knowledge through different subjects.
3. Teachers and students need to participate in green activities regularly after school.
4. Strengthening the relationship between environmental courses and practices.
5. Ensuring government funding and resources support exist.

13.4 Tactics to the Improvement of EE in Chinese Schools (and in Other Similar Countries)?

The results of this study indicate EE in Chinese elementary and secondary schools has achieved progress, yet there are some barriers to the further improvement of EE. In order for Chinese elementary and secondary schools to implement EE more effectively, several changes must be made; otherwise, EE will remain an unimportant concept occasionally included in the school curricula.

13.4.1 Training of Teachers

Teacher plays a leading role in introducing/instructing EE in elementary and secondary schools. The EE knowledge and understanding of a teacher are critical to the realization of EE goals for our students. Our study shows that most teachers found themselves inadequately prepared to teach EE in their classrooms; therefore there is an urgent need to promote training of teachers to better prepare them to teach EE in their classrooms.

Teachers who have received preservice or in-service training spend significantly more class time devoted to EE than those who have not received any training (Lane et al. 1995). There is no doubt that even the best curricula and textbooks are no substitute for teachers who fully understand the goals and objectives of EE. Hence, both preservice and in-service training programs should be carried out for teachers and educational administrations based on some of the strategies listed below:

– Although the government has issued a series of documents about the benefits of implementing EE in elementary and secondary schools, teachers rarely read them. Thus, the training of teachers should help to facilitate teachers' accessibility to these official documents.

- Teachers need to be provided with more opportunities to update their knowledge on environmental-related issues. This could be offered through participation in professional development workshops (Norman and Yeshodhara 2008).
- EE in China is still perceived more as science education than as an interdisciplinary field. Indeed, training for teachers in China has focused on content knowledge alone, which "does not ensure that teachers will have the ability to integrate environmental topics into a wide variety of disciplines" (Carter 2013, p.115). Hence, there should be more emphasis on environmental topics for teachers of social studies to help them improve their understanding of implementing EE in an integrated fashion.

The training of teachers is of great importance to promote EE in Chinese elementary and secondary schools. Hence, more time and effort are needed to better prepare/qualify teachers in this area.

13.4.2 Approaches to EE

The government has been promoting the integration of EE into the curriculum of various subjects and encouraging the implementation of the content and topics of EE through various activities. However, according to our study, most teachers mainly implemented EE through discussion by examining pictures and watching videos on environmental issues. It appears that such an approach has failed to effectively improve students' environmental knowledge and attitudes. Therefore, it is important to develop teachers' ability to adopt inquiry-based approaches that involve "investigation of differing viewpoints and value positions, discussions and debates" (Hanchet 2010, p. 107) in classrooms as well as provide opportunities for observation, experiments, and field trips in order to foster a "shift from a primarily passive approach towards a more interactive teaching strategies" (Hanchet 2010, p. 105). In addition, to increase the knowledge and attitudes of students, the approach to teaching must place greater emphasis on three levels of activities (Hines et al. 1986):

- The content of EE activities must provide students with the ecological knowledge that will allow them to make ecological sound decisions with respect to environmental issues.
- A conceptual awareness should be raised with regard to how individual and collective behaviors influence the relationship between quality of life and quality of environment.
- Development of action plans by students to provide them with the opportunity to implement those plans, if so they choose.

13.4.3 Strengthening Government Support

Our study indicates that a better teaching environment for EE in elementary and secondary schools requires further support from the government. In the review of the development of EE in Chinese elementary and secondary schools, it is apparent that the MOE has been playing a leading role in promoting EE in the basic education system. To further improve the development and generalization of EE in the context of elementary and secondary schools, the MOE should promote the continuous inclusion and emphasis of EE in the national curriculum. However, EE initiatives ought not to rely solely on the MOE-led initiatives. Indeed, governments at all levels can also play their prominent roles in making EE more successful (Yusof 1999). Possible actions from the various levels of governments include:

- Formulating laws and regulations on EE in elementary and secondary schools that explicitly indicate the goals, objectives, principles, teaching content, training for educators, teaching equipment and facilities, responsibilities of the administrations, funding, etc.
- Providing increasing financial support, which could be used to facilitate knowledge mobility between scientists and school teachers
- Working in collaboration with environmental organizations and teachers' associations to create guidelines for teacher training programs in EE
- Promoting EE publicity for the general public, which would help to create a sound EE-facilitating environment

13.4.4 Evaluation of School Performance

Our study indicates that the standardization of school-based EE calls for developing a more comprehensive evaluation system on the implementation of EE in schools. This evaluation system could include the following components: analysis of the level of organization and management of EE programs, frequency and quality of in-class and extracurricular classes offered, and publication of material resources based on the results obtained from the previous categories.

13.4.5 Evaluation of Student Performance

Our study shows that despite years of efforts to foster a shift away from the long-existing testing-oriented education system, exams are still the major evaluation form for assessing the performance of students in Chinese elementary and secondary schools. It is obvious that tests alone focusing on students' memorization of environmental knowledge could never serve as a good way to evaluate students' environmental performance including attitude and perception. A more effective evaluation system for EE needs to be developed. We would propose some of the evaluation methods to couple with traditional tests as listed below:

- Environmental projects could be employed to evaluate comprehensiveness and depth of the research and discussion, credibility of the arguments presented, and students' demonstrated proficiency in finding and using appropriate information.
- Writing assessments could be used to evaluate organization, accuracy, clarity, logic, and degree of understanding reflected.

The government and schools should work together to develop a more effective evaluation system to evaluate students' environmental performance around attitude, knowledge, and skill.

It is hoped that future reports will contain outstanding examples that reflect China's ongoing and continuous commitment of EE in the educational system as well as efforts and initiatives made by China to the world environment improvement and its own national sustainable development. More importantly, we believe that other countries can benefit from the discussion taken up here.

Questions

Discussion questions that could help instructors and students to engage in meaningful conversation about the ideas presented in this chapter.

1. Describe how the environmental education policies in China have developed since the 1970s and how they compare to policy development in your country
2. Discuss with your classmates how environmental education in Chinese elementary and secondary schools is similar to and differs from that in your country
3. Considering environmental education programmes in your country, provide two suggestions on improving environmental education programmes in Chinese elementary and secondary schools. Discuss with your classmates the challenges of implementing your suggestions
4. The five objectives of environmental education as outlined in the Tbilisi Declaration were provided in this chapter. Examine each objective and discuss how it has been implemented in your country. Based on your responses, summarize where you think the Tbilisi Declaration should go 40 plus years later

References

Carter, J. (2013). *The status of environmental education in Illinois public high school science and social studies classrooms.* (Dissertation, Department of Literacy Education. Northern Illinois University). Retrieved from http://search.proquest.com/openview/7bf603bb179a78427c9ecab5f30c7993/1?pq-origsite=gscholar&cbl=18750&diss=y

China Daily. (2007). Scientific outlook on development. Retrieved from http://www.chinadaily.com.cn/hqzg/2007–10/31/content_6220586.htm

Han, Q. Q. (2015). Education for sustainable development and climate change education in China: A status report. *Journal of Education for Sustainable Development, 9*(1), 62–77.

Hanchet, S. (2010). *Experimental education in Canadian teacher education*. Montreal: Concordia University.
Hines, J., Hungerford, H., & Tomera, A. (1986). Analysis and synthesis of research on responsible environmental behavior: A Meta-analysis. *The Journal of Environmental Education, 18*(2), 1–8.
Lane, J., Wilke, R., Champeau, R., & Sivek, D. (1995). Strengths and weaknesses of teacher environmental education preparation in Wisconsin. *The Journal of Environmental Education, 27*(1), 36–45.
Li, Q. M. (2003). *Reflections on environmental education in elementary and secondary schools and its developing strategies*. (Master thesis, Guangxi Normal University). Guilin, P. R. China.
Li, J. S. (2004). *Research on the theoretical system of environmental education and case studies on its implementation*. (Master thesis, Nanjing Normal University). Nanjing, P. R. China.
Maryam, L., & Yeshodhara, K. (2008). An empirical study of environmental attitude among higher primary school teachers of India and Iran. *Journal of Human Ecology, 24*(3), 195–200.
Ministry of Education of the People's Republic of China. (2003). *Guidelines for implementing environmental education in elementary and secondary schools by MOE*. Retrieved from http://www.zhb.gov.cn/download/sszn.pdf?COLLCC=1910316161&
Ministry of Environmental of the People's Republic of China. (2011). *National environmental publicity and education program (2011–2015)*. Retrieved from http://www.zhb.gov.cn/gkml/hbb/bwj/201105/t20110506_210316.htm?COLLCC=1910324729&
Norman, N., Jennings, A., & Wahl, L. (2006). *The impact of environmentally-related education on academic achievement: A literature survey*. Retrieved from http://www.crscience.org
Standing Committee of the National People's Congress. (1979). *Environmental protection law of the people's Republic of China. (1979)*. Retrieved from http://baike.haosou.com/doc/7922302-8196457.html
State Council the People's Republic of China. (1994). *China's Agenda 21: White Paper on China's population, environment, and development in the 21st century*. Beijing, China. Retrieved from http://www.acca21.org.cn/ca21pa.html
State Environmental Protection Administration of the People's Republic of China. (1996). *National environmental publicity and education program (1996–2010)*. Retrieved from http://baike.haosou.com/doc/6828877-7046072.html
Tang, X. Q. (1994, July). China's future development strategy–21st century agenda: The 21st century agenda passed by the United Nations environment & development conference. *China's Foreign Trade, 179*, 1–10.
The Belgrade Charter: A global framework for environmental education. (1976). Connect: UNESCO–UNEP Environmental Education Newsletter, 1(1), 1–9.
The Tbilisi Declaration. (1978). Connect: UNESCO–UNEP Environmental Education Newsletter, 7, 3–4.
Yusof, E. (1999). The effects of the Malaysian development of wildlife and national park environmental education program on the environmental knowledge and attitudes of 13–17 year old students. (Doctoral Dissertation). Retrieved from ProQuest dissertations publishing. (3002665).

Part IV
Environmental Education Research and Poetry

Chapter 14
An Improbable International Collaboration: Finding Common Ground

Astrid Steele and Wafaa Mohammed Moawad Abd–El–Aal

Abstract Consider this a story about scholarly collaboration with a twist. The scholars live and work on opposite sides of the globe – one in Canada, the other in Egypt – and they have only ever communicated via digital media. Nonetheless, they have discovered a shared passion that fuels their work: to prepare teacher candidates to bring curiosity, wonder, dedication and a sense of caring for people and for natural environments into their classrooms. Presented in three parts, the authors share the circumstances and development of their initial collaborative environmental education project, and they reflect on the benefits and obstacles of their continued partnership. Finally, they discuss how they each interpret environmental education in their two different educational spheres.

14.1 Finding Common Ground

This a story about scholarly collaboration with a twist – the scholars live and work on opposite sides of the planet and have only ever communicated via digital media. Nonetheless, they have discovered a shared passion that fuels their work: to promote environmental and sustainability education and action amongst their teacher candidates and thereby to prepare them to bring curiosity, wonder, dedication and a sense of caring for people and for natural environments, into their classrooms.

The necessity and ability to find common ground for successful international collaboration have required considerable effort. As Lucie Sauvé (2009) points out, “We are embodied, localized, contextually grounded beings” (p. 325). We understand the world as we live it – different localities and contexts can create barriers that isolate us from each other. Thus, while a common interest in environmental

A. Steele (✉)
Nipissing University, 100 College Drive, Box 5002, North Bay, ON P1B 8L7, Canada
e-mail: astrids@nipissingu.ca

W.M.M. Abd–El–Aal
Beni–Suef University, Beni-Suef, Egypt
e-mail: wafaa.moawad@edu.bsu.edu.eg

G. Reis, J. Scott (eds.), *International Perspectives on the Theory and Practice of Environmental Education: A Reader*, Environmental Discourses in Science Education 3, https://doi.org/10.1007/978-3-319-67732-3_14

education (EE) connected us, it was only through deliberate communications that the connection could become strong enough to support collaboration between two people who live and work in different worlds. We shared specific descriptions of the courses that we teach, the nature of our working circumstances and the nature of our teacher candidates. We reflected on the merits of various research methodologies and theoretical frameworks.

For most of our chapter, we have maintained our individual voices by telling the same story from two vantage points. It has been our ongoing challenge to find common ground (collegially, academically and personally) as we explored our perspectives on environmental and sustainability education. Thus, we invite the readers to find their own points of intersection with our story. Perhaps the recounting of our experiences will inform and inspire environmental educators, regardless of where they find themselves on the planet.

In *Part One: Finding Common Ground*, we recount our initial contact and conversations based on email threads. We go on to describe the development of a research and writing partnership that focused on the work of teacher candidates in an EE project at Beni-Suef University in Egypt. The project resulted in a coauthored paper (Abd–El–Aal and Steele 2013a) and two international conference presentations (Abd–El–Aal and Steele 2014, 2013b). In *Part Two: Reflecting on our Collaboration*, we consider our motivation to work together, describing both the obstacles and the benefits of our partnership. Finally, in *Part Three: EE Lenses*, we examine our differing perceptions of what EE means to us, given our very different life and work circumstances in two countries that vary considerably in their politics and culture.

14.1.1 Part 1: Finding Common Ground

Astrid Upon completing and defending my dissertation in November of 2010, I was excited to have my very first article published in the International Journal of Environmental and Science Education (Steele 2011). It was based on the environmental/science education action research project I had completed with secondary science teachers in Northern Ontario high schools. I was a freshly minted academic, with decades of public education behind me in wide variety of subject areas and grades, invariably embedding outdoor and EE wherever I found myself. In the fall of 2011, I had just begun to work at a small Ontario university as a science education methodology professor and found myself on a steep learning curve in the world of academia.

Wafaa The 5 years I stayed in the UK to get both my Master of Research and Ph.D. gave me the opportunity to investigate how teacher educators in a developed country apply teaching theories into practice in a way that I had never seen in Egypt. After I came back to my home in January 2008, I decided to implement what I had learnt while away to my own faculty of education at the Beni-Suef University. In trying to find a theory that maps and explains EE, I did some research through international journals and found that my work could be theoretically framed by critical place-based pedagogy (Gruenewald 2008). Then, I contacted Astrid (via email) after I read about her project and the role she played for 5 months as a facilitator in order to assist secondary science teachers who had limited EE exposure to embed it into their science curricula. I found a similarity between Astrid's action research and what I was doing at my faculty, as I too support my students to practise EE rather than just reading about it. We both (a) focus on EE professional development, and the challenges that teachers face in practice, and (b) adopt very similar methodologies (action research, with data collection through interviews, meetings, artefacts, etc.). These similarities encouraged me to email Astrid without any hesitation.

Astrid It was without much context but with considerable curiosity that I responded to an email from a professor in Egypt commenting on my article and requesting that I collaborate on her research. Egypt! What an amazing world I had entered – where I might be drawn into a conversation about EE by someone on the other side of the planet. I wanted to know more about Wafaa the person, about her university and her EE course, more about the country of Egypt. (I remembered vaguely that there had been some recent political upheaval.) So, I began an email thread that built a foundation for collaboration:

Dear Prof. Abd–El–Aal,

Thank you so much for contacting me. It is very affirming for me that the work that I did here in Canada is being considered by scholars in other countries. I would be most interested in starting a dialogue with you. I would like to hear more about the course(s) that you are teaching, a description of your students and the type of action research you might be interested in, and perhaps some of the other work/research that you are doing. (5/19/11)

Wafaa I asked Astrid to co-write a research paper on the EE course I had developed for teacher candidates. After sending her the data (comprised of videos, photos and text), transcribed and translated from Arabic into English, we started to organize our paper and take individual and cooperative tasks. There was no consideration for any ethnic issues between us. One of the funny things in this challenging collaboration was that I first did not know Astrid's gender and I did not mind, which conflicts with other Arab women's views. This is because I was looking for a person to share my perspective and help me explain my view on EE, regardless of gender.

Astrid The response that I received set me on a journey of re-understanding and re-imagining what EE might encompass, beyond my privileged perception rooted in the Western world. I had no conception of what EE might look in a majority world country, indeed had not ever given it much thought. (Even the term *majority* world was new for me – a descriptor without the colonial implications inherent in the term *developing world*. *Majority world* highlights that the majority of humankind lives in the countries of Asia, Africa and Latin America and points out the incongruity that the group of eight countries – whose decisions affect majority of the world's peoples – represent a tiny fraction of humankind [as defined by Appropedia]). Wafaa wrote with passion about her desire to introduce her teacher candidates to nontraditional EE studies:

Dear Prof. Steele,

... environmental education courses in Egypt are grounded on general environmental knowledge that mostly talks about topics such as different kinds of pollutions. As I mentioned before, student teachers are neither encouraged to investigate their own local environmental problems nor reflect on such problems. They are taught some environmental topics in closed classrooms without any opportunity for students to do fieldwork. Also, Egyptian research on environmental education reflects Egyptian researchers' traditional view of learning – rote learning, i.e. learning for memorizing massive knowledge about environment, which some Egyptian researchers call environmental literacy! Thus, designing an environmental education course based on fieldwork, students' taking part in solving what their local areas suffer from would be a NEW trend in Egypt. It is worth/interesting to study the impact that the January Revolution had in Egypt and how it has affected my students and their desire to put their environmental learning into action, i.e. environmental citizenship. I would call this study, Towards Environmental Citizenship in Teacher Education Programmes in Egypt. (5/31/11)

Astrid Initially, when I agreed to collaborate with Wafaa in writing a paper about place-based EE in an Egyptian faculty of education, I anticipated that it would be an opportunity to compare what I thought of as advanced EE methods found in Canada (such as activity-based student-focused inquiry) with the fledgling efforts in Egypt, but that would not be the case. Wafaa had already implemented those pedagogies and had collected significant data, in the form of transcripts of interviews/fieldwork (in Arabic), photos and videos, based on EE projects that her teacher candidates had undertaken. The projects themselves were compelling, including meeting with shopkeepers to discourage the use of precious water to wash their sidewalks, meeting with a town council to encourage and support the clean-up of a local water way and modelling and encouraging a litter clean-up in a neighbourhood. As I read the stories of the Egyptian students working in their communities to clean up, rebuild and educate community members in how to care for their environments, the concept of place-based education took on new meaning. Not only were the students learning about their communities but they were working to decolonize those communities, to create a new vision of how their communities might function.

Wafaa Trying to accomplish the aims of the EE course I was teaching, I thought about the relationship between what a student studies and the place/communities in which they live.

Dear Astrid

Please find attached an interesting file (Fig. 14.1) indicating the impact of my environmental education course as well as the Egyptian revolution on the students' environmental citizenship. Although it's in Arabic, you would see how students who called themselves Hope Group, and called their problem the Disaster, determined their own environmental problem in their village and conducted interviews with people there not only to reflect on

Fig. 14.1 Hope Group's focus area for water and garbage clean-up in their village

their unacceptable environmental behaviour, but also to solve this problem. The students tried to convince those people to co-operate with each other to maintain the water in their village clean. They also encouraged the people to make a complaint to their local council. Actually, I'm so proud of my students, the CDs they submitted to me contain lots of data which reflects a beginning of awareness of their roles and rights. Regards, Wafaa. (6/4/11)

Astrid Wafaa's request was that I assist her in analysing the data and writing about it. This required that she translated the transcripts from Arabic to English so that I could read them – an enormous undertaking. Within a week, English transcripts began to appear in my inbox, and I was captivated by the narratives of impassioned young Egyptian teacher candidates striving to improve the well-being of their communities. The recent Arab Spring figured notably in their discourse: the desire of ordinary Egyptian people to create a better life for themselves was evident in the many references to environmental citizenship. (By then I had done my own research to understand contemporary Egypt, beyond my romanticized images of pharaohs, pyramids and camels.) I found myself wondering if the teacher candidates at my institution could be equally motivated to make such significant contributions to their own communities. Certainly, some of them had already done so, but for the most part, teacher candidates might speak about the importance of healthy communities although many are complacent in their actions. One needs look no further than the waste bins at the university, many filled to overflowing with plastic and non-recycled items. I began to question if what I thought of as a Canadian version of EE, with outdoor field schools, recycling projects and EE-enhanced curriculum,

Fig. 14.2 Sample slides of the presentation at WEEC 2013, in English and Arabic

> was all that effective. Perhaps I had a few things to learn from my Egyptian colleague. Preparing our manuscript for publication was accomplished over several months of emails, during which I continued to struggle with how I could contribute to Wafaa's research project. I wondered if it was even appropriate for me, with my Western views, to analyse a project carried out in an eastern majority world country.

Dear Wafaa,

I have read a number of Dr. Greenwood's articles (he formerly went by the name of Gruenewald) in which he theorizes the connection between place-based education, which is essentially an integrated environmental study of a specific place – and critical pedagogy that seeks to identify and rectify social justice issues. Greenwood identifies two concepts: decolonization and re-inhabitation; I especially find the concept of re-inhabitation compelling, particularly given the example in the transcript in which the students, the farmer and the woman try to come to terms with how best to solve the water pollution problem. I also wonder if the themes of decolonization and re-inhabitation might speak clearly to the changes that you described in your email, of the attitudes of students and others... a sense of liberation and that there is an ability to move forward into a better life. (6/20/11)

Wafaa's reply came the following day:

Dear Astrid,

Thanks for sending this article (Gruenewald 2008: Critical Pedagogy of Place), it really touches our research. The rest of data also confirms your view as well. … I agree with you that the themes of "decolonization" and "re-inhabitation" describe what my students did in their communities. (6/21/11)

Critical place-based pedagogy provided a framework for analysing the videos, the photos and the translated conversations between fervent EE students and villagers, shopkeepers, students, farmers and apartment dwellers. As our collaboration continued, the seeds of friendship were sown, and I became increasingly interested in finding an opportunity to visit Wafaa in person. When a call for papers came from the World Environmental Education Congress WEEC 2013, to be held in Marrakesh, Morocco, I thought perhaps this would be the opportunity to finally meet Wafaa and present our paper together at the conference!

Wafaa We both worked to prepare the research paper for publication in the *Journal of Education for Sustainable Development* (JESD) (Abd–El–Aal and Steele 2013a) as well as for the presentation at the WEEC 2013 in Morocco, which had to be written in Arabic and English, a requirement from the conference organizers (Fig. 14.2). This took many emails in which we both tried to explain our views on the paper analysis. I got so excited to meet Astrid at that conference and present our paper there. However, due to a financial constraint, I was neither able to attend the conference nor meet Astrid face to face! Later, Astrid let me know what the conference attendants' comments on our paper were and their questions. Although I felt disappointed, until now, I have never thought that I wouldn't meet Astrid one day!

Astrid Alone, I travelled to northern Africa to make the presentation for both of us, wondering at the professional funding inequity that would enable one scholar to travel halfway around the world while the other did not have the professional funds to travel on the same continent. Indeed, our next paper was to be presented by Wafaa at a conference in Spain, yet again her attempts to attend were stymied by bureaucracy. As a scholar, Wafaa faces barriers of economy and bureaucracy that I do not. Yet, she carried on with her research and soon suggested a second project, this time addressing students' understanding of science content.

Wafaa After I investigated my students' science content knowledge, I again asked Astrid to co-write a paper to present at EDULEARN14 in Spain. Once again, we worked together via emails in order to prepare the paper for this conference. She was not able to attend the conference, but I did my best to do so. I had to submit many certificates to the Spanish embassy, which took much time and travel to Cairo. Unfortunately, I neither presented the paper at that conference nor met Astrid.

Dear Astrid,

For the visa, I am so disappointed and frustrated. You cannot imagine the number of documents required! For example, my marriage certificate, local ID, HR letter, accommodating reservation, ticket/s, bank statement, 4 pages– application form etc. All those have to be original and a translated copy of them. …You cannot imagine what I am feeling now! I was so happy when the abstract was accepted, but now I am... Anyway, as we believe here everything is fated. (5/15/14)

Astrid Despite setbacks, our collaborative projects continued, branching into other intersections of science and EE including the use of social media and socio-environmental issue-based pedagogies. We took this experience of writing about our international collaboration as an opportunity for reflection, as a gateway to moving forward with deeper understanding of our work. In the next section, we give thought to our shared work and experiences.

14.1.2 *Part 2: Reflecting on Our Collaboration*

Wafaa No doubt that my collaboration with Astrid, which helped me find a theory (place-based education) that supports and explains my work in EE, encouraged me to ask for more cooperation. The applause that our first paper received at WEEC 2013, as well as the JESD reviewers' comments, motivated me to ask her more about how a specific course such as the science education course is taught at her faculty. Then, I thought about the ways in which I, and my students, could benefit from a redevelopment of the course I teach. Astrid was helpful with advice and comments on other research projects I had conducted; she provided website addresses and journal articles that are not available through my institution. However, in spite of the benefits, there were some obstacles, for example, sometimes she was busy conducting research projects or was in the field with her students so communication was delayed. Also, she is off work on Saturdays and Sundays, but I work on these days so communicating could be troublesome. Sometimes I found connectivity with the web at my faculty and at home unavailable or too slow to send emails or attach files, especially videos and photos.

Astrid Throughout our extensive correspondence of over 5 years, we have never actually met nor have we Skyped (initially, Wafaa's Internet connections were not reliable) and so have used email exclusively. Despite what might appear to some to be a limited form of communication, we have come to know each other, sharing our personal lives and professional endeavours. Language is not a barrier as much as an encumbrance, particularly for Wafaa who is faced with communicating in a second language and not infrequently must translate Arabic data for me. So we have tried to make our words count. We encountered cultural stumbling blocks, for example, the differences between the two university cultures in terms of funding and grant resources, collegial support and scholarly expectations and the difference in ease of travel to attend conferences. The improbability of our international collaboration lies within the generally recognized requirements for collaboration, which are a shared purpose, collegial relations, norms of improvement and structures (Wallace et al. 2007). We share a common purpose (i.e. to provide environmental and sustainability education for our teacher candidates), we have developed a collegial relationship, and we both seek to improve our practice. However, the structures that would allow us to share time, space and resources have proven to be inhibiting and have required persistence and patience to find workable solutions.

Wafaa This points to the importance of communication between teacher educators in all parts of the world. Our cooperation reflects a kind of integrating view about the environment and our roles as teacher educators towards it. Emails played the main role in our contact. Sometimes it took me many hours a day to send Astrid the videos the students took during their fieldwork. This highlights the importance of adequate communication technologies between researchers in different countries.

Astrid I know that our collaboration has enriched me in numerous ways and not only professionally. Over the years, Wafaa and I have shared bits of our personal lives as well: health concerns of our families, concern for a changing climate and imagining what it might be like to visit each other. Simple conversations nevertheless have had profound impacts as collaboration is relational and requires strengthening not only of professional bonds but of personal understandings as well. Based on a growing understanding of the cultural, geographical and political spaces that we individually inhabit, we have had to invent a unique shared global space for ourselves. Noel Gough (2013) describes this as "a process of constructing transcultural 'spaces' in which scholars from different localities collaborate in reframing and in decentering their own knowledge traditions and negotiate trust in each other's contributions to their collective work" (p. 41). Our communications, and the resulting collaboration, bear significant semblance to what Michael Peters and Daniel Araya (2009) describe as a peer-to-peer (P2P) learning network or ecology. Of course ours is a very small network with only two nodes of intersection; still, our "voluntary collaboration between two equipotent partners" (p. 246) resonates with many of the criteria set for the P2P learning ecology: (a) informal and responsive, (b) consistently evolving over time in directions that we choose, (c) embracing a growing level of trust, (d) decentralized and participatory and (e) tolerating experimentation and failure. Recognizing that our collaboration has created opportunities for learning and professional development for both of us, we continue to work within, and hopefully nurture, our P2P network.

14.1.3 Part 3: EE Lenses

In this final section, we share our thinking about EE, again from our two perspectives. Wafaa discusses the difficulties of providing what she views as meaningful EE experiences in an Egyptian context while Astrid compares the efficacy of a Western rendering of EE research with the work that Wafaa is doing in Egypt.

Wafaa When I looked at the EE books my colleagues wrote and utilized with students, I found that they focused on giving the students general information about the environment and associated problems such as pollution but without any activities that would help them interact directly with their local environments. I dislike studying or teaching only abstract theories without practicum and activities. Unfortunately, I was faced with obstacles, including the lack of resources and a misunderstanding by my colleagues of the "new teaching methodology" I brought from the UK. My colleagues felt that "I had bought some materials and tools in order to play with the students instead of teaching the science content that the students have to learn". Although this upsets me, I tried to move forward to achieve my own views about teaching and the relationship between the teacher and students. I consider my colleagues' misunderstanding/unawareness regarding my students' fieldwork and practicum as one of the main barriers to benefiting from Western perspectives. They have never done what I do. Their comments "What are you doing? You're now here in Egypt, not in the UK." are wearying for me. Although there is a consensus in the scholarly literature on the importance of EE in developing students' attitudes towards environment in Egypt (see, e.g., Al–Tanawey and Al–Sherbeiney 1998), they have not taken the opportunity to participate in solving any environmental problems in their communities. As environmental degradation is one of the serious threats facing Egypt, there has been an increase in EE at all levels of schooling. For example, in the primary school, Arabic language curricula address environmental issues (e.g. keeping streets clean from garbage, maintaining environmental resources, etc.). At the university level, at Beni-Suef University, students in the preschool department have a mandatory EE course. Since the 1970s teacher educators have conducted many studies on EE, resulting in the development of EE courses (e.g. Abd–El–Salam 1991). The EE courses and research in Egypt have focused mainly on quantity rather than quality, typically providing content knowledge and employing quantitative questionnaire methods. But, these methods do not reveal or impact what people actually do in the places they inhabit. If we want to change attitudes and behaviours, we must adopt a different pedagogy that helps learners think critically about actual environmental issues surrounding them, allowing them to interact with their places as effective citizens who can make wise decisions regarding the present and with an eye to the future. This "critical pedagogy of place challenges all educators to reflect on the relationship between the kind of education they pursue and the kind of places we inhabit and leave behind for future generations" (Gruenewald 2008, p. 308). In Egypt, unless there is a united view/strategy of implementing EE in schools and universities, what I do and aim to achieve would be merely an individual effort. International cooperation in the field of environment is important, as a way for EE researchers and teachers to exchange their views on sustainable development through different educational lenses.

Astrid As I pondered our work with teacher candidates, I wondered if EE really could fulfil its promise to transcend cultures and exist as a global enterprise (UNESCO 2014). In the *International Handbook of Research on Environmental Education* (2013), Robert Stevenson, Arjen Wals, Justin Dillon and Michael Brody emphasize five characteristics that identify EE: (1) a normative or value-laden endeavour; (2) interdisciplinary, embracing environmental, sociocultural and economic elements; (3) not only the development of knowledge, skills and attitudes but also agency to advocate and act for change; (4) enacted in both formal and informal spaces; and (5) local and global in scope. Each of the five characteristics resonated with the EE research and teaching that I have been doing in Canada, but I wondered if they would be adequate to the work that Wafaa has been doing in Egypt. In other words, I was asking if the perspectives purported by four highly respected EE scholars from the Western world would ring true in Egypt, a majority world country. From my understanding of her work, Wafaa does indeed encourage her students to enact change in local, informal spaces in their communities; she recognizes the impact of political, social and economic factors that play into the interactions between her students and community members.

14.2 And Finally...

Wafaa If we want to participate effectively in our communities and protect ourselves and the places we live, we can benefit from others' experiences on an international stage. This does not mean that educational systems in majority world countries adopt any methodology or pedagogy. The criteria that would govern what majority world countries will import are dependent not only on culture and economic differences but also on the nature of the environmental issues in every country. Developing an international partnership for EE that aims at encouraging environmental protection, rationing environmental resources, recycling mechanism, etc. can shorten the distance the majority world countries would take to solve their environmental problems. I consider my collaboration with Astrid a kind of distance education through which two teacher educators in the field of EE, in two very different countries, provide an example of how we can together promote sustainable EE practices in teacher education programmes.

Astrid As an educator in Canada, I thought that I could offer insights to those educating for environment in the majority world, but experience demonstrates that I have much to learn about what EE really means globally and how small my own offerings are in comparison to those who are most impacted by environmental degradation. I want to explore what it means to be an environmental educator when confronted with the realities of the majority world. I know that my reflections will continue to be shaped by my Western thinking pathways, but I also recognize that my collaborations with Wafaa have informed and broadened that thinking. Paul Hart (2005) provides a bridge of sorts to mitigate my tensions and the differences between Wafaa's and my perceptions of what we might offer each other: "All we are doing, as reflexive researchers, is to write in ways that reveal the limits of our knowledge, our political orientation and other dimensions of self, in ways that reveal the discourses that shape our work and open possibilities for thinking about our work as we get on with it" (p. 399). Indeed, we look forward to getting on with our work within the global EE commons.

Questions

Discussion questions that could help instructors and students to engage in meaningful conversation about the ideas presented in this chapter.

1. How might the political climate in Egypt during the Arab Spring (2011) have influenced the direction of the research and the actions of the teachers-in-training?
2. As you listen to the voices of both Wafaa and Astrid, do significant differences and/or commonalities in culture emerge? If so, comment on how those might have influenced the collaboration.
3. How do you see this collaboration of international researchers benefiting each author and their students?
4. Comment on the authors' strong desire to meet face-to-face. Why does this seem so important? What might be accomplished by such a visit?

References

Abd–El–Aal, W. M. M., & Steele, A. (2013a). Practicing environmental citizenship in Egypt: Hopes and challenges encountered. *Journal of Education for Sustainable Development, 7*(2), 1–22. https://doi.org/10.1177/0973408214526489.

Abd–El–Aal, W.M.M., & Steele, A. (2013b). *Towards practicing environmental citizenship in Egypt: Hopes and challenges*. World Environmental Education Congress, Marrakesh, Morocco.

Abd–El–Aal, W.M.M., & Steele, A. (2014). Pre–service teachers' science content knowledge: An Egyptian perspective. *EDULEARN14 Proceedings* (pp. 4407–4417), Barcelona, Spain.

Abd–El–Salam, A. M. (1991). *Environmental culture of students at Mansoura University* (pp. 113–146). Paper presented at the third scientific conference of the Egyptian association for curricula and methodologies: Future view on curricula in Arab world, 4–8 August, Alexandria, Egypt.

Al–Tanawey, E. M., & Al–Sherbeiney, F. A. (1998). The effect of a proposed programme based on self–education in environmental education for faculty of education students on developing their environmental awareness and attitudes. *Journal of Science Education, Egyptian Association for Science Education, 1*(2), 21–73. (in Arabic).

Gough, N. (2013). Thinking globally in environmental education: A critical history. In R. B. Stevenson, M. Brody, J. Dillon, & A. E. J. Wals (Eds.), *The international handbook of research on environmental education* (pp. 33–44). New York: Routledge. https://doi.org/10.4324/9780203813331.ch4.

Gruenewald, D. A. (2008). The best of both worlds: A critical pedagogy of place. *Environmental Education Research, 14*(3), 308–324. https://doi.org/10.3102/0013189X032004003.

Hart, P. (2005). Transitions in thought and practice: Links, divergences and contradictions in post–critical inquiry. *Environmental Education Research, 11*(4), 391–400. https://doi.org/10.1080/13504620500169270.

Ismaeil, H. H. S. (2004). *Environmental values and behaviours of primary science teachers: Evaluation study* (Unpublished M. Ed. thesis). Cairo: Ain Shams University, Faculty of Education (in Arabic).

Othman, A. K. M., Al–Magrabey, R. M. M., & Hajjaj, A. A. (2012). Kindergarten teachers' awareness of some technological environmental problems. *Journal of Al–Fath, 48*(1), 68–92. (in Arabic).

Peters, M. A., & Araya, D. (2009). Network logic: An ecological approach to knowledge and learning. In M. McKenzie, P. Hart, H. Bai, & B. Jickling (Eds.), *Fields of green: Restorying culture, environment and education* (pp. 239–250). Cresskill: Hampton Press.

Sauvé, L. (2009). Being here together. In M. McKenzie, P. Hart, H. Bai, & B. Jickling (Eds.), *Fields of green: Restorying culture, environment and education* (pp. 325–336). Cresskill: Hampton Press.

Steele, A. (2011). Beyond contradiction: Exploring the work of secondary science teachers as they embed environmental education in curricula. *International Journal of Environmental and Science Education., 6*(1), 1–22.

Stevenson, R. B., Wals, A. E. J., Dillon, J., & Brody, M. (2013). Introduction: An orientation to environmental education and the handbook. In R. B. Stevenson, M. Brody, J. Dillon, & A. E. J. Wals (Eds.), *The international handbook of research on environmental education* (pp. 1–6). New York: Routledge. https://doi.org/10.4324/9780203813331.ch1.

United Nations Educational, Scientific and Cultural Organization. (2014). *Shaping the future we want: UN decade of education for sustainable development.* Paris: UNESCO. Retrieved from http://unesdoc.unesco.org/images/0023/002301/230171e.pdf.

Wallace, J., Sheffield, R., Rennie, L., & Venville, G. (2007). Looking back, looking forward: Researching the conditions for curriculum integration in the middle years of schooling. *The Australian Educational Researcher, 34*(2), 29–49. https://doi.org/10.1007/BF03216856.

Chapter 15
Environmental Education Research and the Political Dimension of Education for Citizenship: The Brazilian Context

Luiz Marcelo de Carvalho and Heluane Aparecida Lemos de Souza

Abstract In this chapter, we aim to explore the relationship between education and the process of citizenship construction in Environment Education (EE) research reports and essays (master and doctoral studies) carried out in Brazil, between 1981 and 2012. Considering the texts analyzed as a discursive processes, which take part in an infinite chain of meaning production, our study enables us, in one hand, to recognize how susceptible we are to generalizations and to decontextualized universalizations in which the ideal of citizenship is founded on principles of liberal or neoliberal inspiration. On the other hand, resistance to the construction of false universalities, which are supported by the ideologies of hegemonic groups, was clearly present in a set of these analyzed texts. As part of an international researchers and educators' community, it is important to recognize that we are involved in a context that is full of conflicting ideological interests. This reality, in turn, influences the formulation and the implementation of public policies in education/environmental education as well as knowledge production and dissemination.

15.1 Environmental Education and Citizenship Education

The relationship between education and the processes of citizenship construction has been exhaustively explored and emphasized in the literature (Peixoto 2010). According to Arbués (2014), "since the final decade of the twentieth century

L.M. de Carvalho (✉)
Departamento de Educação. Instituto de Biociências, Universidade Estadual Paulista, Unesp. Campus de Rio Claro. Av. 24A, 1515, Bela Vista, 13506-900 Rio Claro, São Paulo, Brazil
e-mail: lmarcelo@rc.unesp.br

H.A.L. de Souza
Departamento de Educação, Instituto de Biociências, Unesp, Campus de Rio Claro – Av. 24A, 1515, Bela Vista, 13506-690 Rio Claro, São Paulo, Brazil
e-mail: helu_souza@hotmail.com

G. Reis, J. Scott (eds.), *International Perspectives on the Theory and Practice of Environmental Education: A Reader*, Environmental Discourses in Science Education 3, https://doi.org/10.1007/978-3-319-67732-3_15

western democracies have witnessed the ongoing development and consolidation of education for citizenship" (p. 227).

Nowadays, perhaps more so than at any other historical time, educational policies have prioritized the social aspects of education (Arbués 2014). Among these social aspects, the trend of developing and consolidating citizenship education in different countries has been made clear since the end of the last decade of the twentieth century. According to Arbués (2014), the fall of the Berlin wall and the hard-hitting criticism of extreme liberalism, which began at the end of 1980s in the United States and Canada, had repercussions in European countries. Educational orientations in France in 1999, in England in 2002, in Germany in 2004, and in Italy and in Spain – around the same period – can be seen as example of these policies mentioned by Arbués (2014) that have been incorporated by the European Council. According to this author, in these countries, educational policies have both emphasized the personal dimension of human beings, the achievement of individual liberties, and, at the same time, the social aspect of communities.

According to Bannell et al. (2011), beginning in the 1990s, the concept of citizenship became central to both political and educational theories, particularly concerning human rights. Such interest, although visible in different social spheres, has been implemented in different curriculum documents in various countries, such as those already mentioned in Europe, but also in North America (United States and Canada) and South America – see, for example, the Curricular Proposals in Brazil since the end of the 1990s (Bannel et al. 2011).

However, as Bannel et al. (2011) indicate, "the concept of citizenship presupposed in those documents is not explicit" (p. 7) – and the official documents that deal with the objectives and purposes of education in Brazil were not exception. This, in turn, raises doubts concerning the possible relationship between the educational process and citizenship education:

> Why have educational reforms in recent decades placed education for citizenship as a main objective of basic teaching in so many countries, including Brazil? Which concept(s) of citizenship are these reforms based on? Is the new discourse of citizenship an ideology in favour of the dominant classes' interests or is it part of a mobilization in favor of human emancipation? (...) why has the category of citizenship become the main focus of the philosophical liberal policy over the last two decades? (Bannell, Prata and Fenerich 2011, p. 7)

According to Valle (2010, p. 214), the emphasis to relate education and citizenship that could be taken as a "new demand," in fact, correspond "to a real return to the origins" of the educational practices. According to the author, the first and also the most constant meaning attributed to education, since education started to be seen as a social practice, was to educate "future citizens" (p. 214).

Taking into account the Brazilian context, Gallo (2010) considers that the ideal of citizenship "has always been tenuous, riddled with controversies and contradictions" (p. 133). According to the same author, after the country went through two decades of military right-wing dictatorship, "the last decade and a half has been marked by the discourse of the construction of citizenship" (p. 134). In this discourse, education is always remembered as playing a key role in this process. However, in order to make informed choices, coherent with their political orientations, educators in general and researchers in particular should make it clear which citizenship and which education they are talking about.

The community of researchers and educators in EE has not spared themselves from taking part in this debate or from bringing concrete contributions to this meaning-making process in education and citizenship. Environmental education, according to Jacobi (2005), for example, should be placed in a broader context, namely, education for citizenship itself and understood as a practice that is "decisive in the consolidation of citizen–subjects" (p. 243).

Loureiro (2011) understands EE as being a constituent part of the social/environmental movement, which mainly focuses on relationships between human beings (society), nature, and the process of constructing planetary citizenship. To this author, planetary citizenship or ecocitizenship is a new concept:

> [It expresses] the inclusion of ecological ethics and its consequences in everyday life in a context that enables individual and collective awareness of local, community and global responsibilities, having respect for life as a central axis and defending rights to it in a world without geopolitical boundaries. . Doing so, the feeling of belonging to humanity and to a single planet is a key idea In this concept. (Loureiro 2011, p. 80)

In an international context, Scott (2011) explored research studies and general literature regarding school practices and contributions to "young people's learning about living on the planet both sustainably and well" (p. 409). The author concludes that "the priority must be to engage young people with ideas about sustainability through imaginative teaching strategies that provide stimulating opportunities for learning, including practice in **citizenly engagement** – and that everything else has to be secondary to this" (p. 409; bolded emphasis added).

To provide an idea of the centrality of citizen/citizenship in the EE discourse on an international level, a search of articles published since 1997 in the journal *Environmental Education Research* (EER) returned 453 matches. In a Special Issue (SI) recently edited by this journal (volume 21, issue 3), whose theme was explicitly linked to the role of EE regarding the political dimension of education ("environmental education in a neoliberal climate"), the editorial team stressed the relationship between EE and citizenship (Hursh et al. 2015). Therefore, certain papers included in this special issue address the relationship between EE and the construction of citizenship as an intrinsic and inherent function of education itself.

According to Dimick (2015), for example, "developing students' civic capacities and dispositions to engage as participatory citizens in relation to environmental issues and concerns" (p. 390) should be an educational aim of EE. In the same direction, Stahelin et al. (2015) recognize the emphasis that we can find in the Brazilian EE context linking EE "to the formation of citizen–subjects capable of critically reading socio–environmental realities and mobilizing collectively to intervene// in transformative ways to overcome social injustice" (p. 437).

In 2014, the Brazilian Journal of Research in Science Education published a special issue with the title "Environmental discourses in science education: Contributions to democracy, citizenship and social justice." As clearly shown in the title of this special issue, this publication considered citizenship as a key concept involved in EE. According to the editors, "current discourses of democracy, citizenship and social justice are increasingly recognized as being amongst the many existing factors with which education interacts as a means to assist society in the development of scientifically and ecologically literate citizens" (Reis and Oliveira 2014, p. 11). Several papers published in this special issue specifically address the links between EE and citizenship by exploring different pedagogical strategies. For example, Wolmann and Braibante (2014) report on a teaching experience in which they encourage high school students to prepare informative leaflets related to EE. Conrado et al. (2014) explore teaching strategies based on problem-based learning (PBL), which "may contribute to the education of responsible citizens in social and environmental contexts" (p. 80). Moreover, Cosenza et al.'s (2014) research aims to understand how in-service teachers discursively articulate certain science education goals, especially those concerning the relationships between science, society, and citizenship, with environmental justice and risk (p. 90). Finally, through the analysis of similar research reports that have linked EE and science education, Santos et al. (2014) pointed out trends regarding the political dimensions emphasized by the authors in those texts. Although their results have shown that citizenship is one of the most stressed aspects related to the political dimension of EE, they have concluded that there is a need to clarify the meanings of the political dimension of EE in general and the relationship between education and citizenship.

As the articles above indicate, the relationship between education and citizen/citizenship is frequently emphasized in EE research reports and EE theoretical reflections in Brazil as well as at the international level. According to Higuchi and Moreira Júnior (2009), what researchers envision is the possibility of rapprochement between the assumptions that guide the educational practice of EE with research practices.

There is no room for naivety. There are great risks of inconsequent spontaneism and pragmatism when we face the real meaning of the political dimension of EE education, particularly in regard to citizenship. There seems to be no alternative other than attempting to understand such proposition in more depth. As Valle (2010) suggests, the "analysis of the philosophical or historical meanings which can be attributed to the notion of citizenship [is] an essential task for education" (p. 219).

With this regard, it is reasonable to argue that the term "citizenship education" carries a variety of meanings and understandings that depend on a number, including the ontological, epistemological, ideological, and political perspectives of scholars and educators. As Reis and Oliveira (2014) point out, it should be considered that educators, including us, "struggle with competing and ideologically disparate concepts of not only citizenship, but also of democracy" (p. 20). Therefore, we hope that our reflections may take part in that "real task which is the political, collective construction of this meaning" (Valle 2010, p. 219).

The broader issue that we have chosen to explore in this chapter is the role of EE in the citizenship education process in research reports or theoretical essays carried out in Brazil. As we have shared, although this perspective has been emphasized by the Brazilian community of researchers enrolled either in education or in the EE field, it has also been historically considered and explored by international researchers reporting in these areas. The questions guiding our investigation include:

Have environmental education researchers stressed aspects related to the relationship between EE and citizen/citizenship in their investigations? Has the EE research carried out in Brazil been designed in a way that helps to clarify possible relationships between EE and the process of citizenship construction? What aspects related to the EE–citizenship education relationship have been highlighted by researchers?

15.2 The Meaning-Making Process in EE and Citizenship Construction in Brazilian Research Reports

This chapter focuses on attempting to explore meaning-making processes surrounding the relationship between EE and citizenship education analyzing doctoral and master studies (research reports). If we consider those texts as a discursive genre and a discursive practice, they can be seen as constituents of the EE field and influential in the meaning-making process regarding such links between education and citizenship.

We believe that particular Brazilian studies may constitute significant contributions to foster, as proposed by Bakhtin (2010), a dialogical attitude, a relationship with the other. Only this dialogical attitude among researchers and educators all over the world can bring to surface different utterances, which constitutes a link in an endless chain of utterances regarding key concepts to the EE field at the international level.

We have systematized possible meanings attributed to the relationship between EE and citizenship in theses and dissertations on environmental education carried out in Brazil. The documents analyzed were selected from the database organized by the EArte Project team, an interinstitutional research group that has developed a broad research project in the EE field.

In addition to gathering all EE theses and dissertations concluded in Brazil since 1981 – from the first master's EE study produced in the country – at national level and as extensively as possible, the mentioned EArte project aims to organize a database in which all these research are catalogued. Other objective of this research project is to analyze some particularities regarding institutional, regional, and educational contexts, such as universities, postgraduation programs, and Brazilian's regions in which theses research were carried out. The project's website (www.earte.net) gives access to the Thesis and Dissertation Database and offers a panoramic view of the project, including the history, objectives, and the selection and classification criteria used to select the EE theses and dissertations included in the database.

In terms of methodological approach, our starting point is that research reports are not a picture of reality, but rather results of discursive interactions, links in a verbal communication chain, and answers to concrete utterances already produced and that will probably stimulate other utterances (Ferreira 2002). Considering this, we cannot utilize data reported in research as reliable records of reality, but as the results of a look at reality with plenty of intentions (Martins 2006). According to Payne (2009), the aim is that the analyses of EE research can contribute to our understanding of research texts as part of the process of construction of discourses about EE. It is a way of clarifying the meanings that we have constructed about these educational practices and, thus, our understanding of their identities, as a field of educational practices and research.

Considering the abstract of a thesis or a dissertation as an autonomous text and a means with its own particular characteristics or, in other words, a discursive genre (Bakhtin 2010; Ferreira 2002) considers justifiable and appropriate that we take these texts and analyze them in terms of state-of-the-art approaches in a given academic production. Taking abstracts as discursive genres means considering these texts as having been produced for a "determined purpose" and "under specific conditions of production" (p. 267). In Bakhtin's dialogical perspective, each abstract participates in the "verbal communication chain" (Ferreira 2002, p. 270).

Considering these perspectives, we have analyzed the abstracts of the 3180 theses and dissertations that constitute the EArte database, trying to identify units of register and, from them, units of meanings regarding the relationship between EE and the process of citizenship construction.

15.3 The Ideal of Citizenship in Environmental Education: Theses and Dissertations in Brazil

Considering the total number of abstracts of theses and dissertations that constitute the EArte database (3180), 1780 of them (56%) refer to the political dimensions of EE. In addition, approximately 28,5% of those 1780 studies (*n*= 510) use the concept of citizenship to do so. In our analysis, we also tasked ourselves to recognize different nuclei of meanings regarding the relationship between EE and citizenship education from the analysis of the titles and abstracts of the theses and dissertations catalogued on the database of our choice (Table 15.1).

This panoramic framework shows that in 48.5% (*n*= 248) of the studies in which the term citizenship is found, its relationship with EE is assumed a priori. That is, there is no evidence that this issue has been problematized or critically studied in depth (Group A, Table 15.1).

In Subgroup A1 (35,5% of the total or *n*= 182), the term citizenship is incorporated into the text as a discursive resource to qualify the individual. In these research reports, the term citizen is used to refer to an individual or a group of individuals, such as "the students as a group of citizens." Taking this data into account, it seems there is no deliberate intention to politicize the EE discourse. In this case, the concept of citizenship or citizen is brought into the text to emphasize that individuals have the citizen status or are about to achieve that condition; citizenship is considered to be a result of educational practice, whether in a school context or not.

Those studies in Subgroup A2 (approximately 13% or *n*= 66) also presented the concepts of citizenship or citizens as resources to qualify an individual's educational process. We found pre-indicators in those texts that referred the reader to political perspectives of EE, taking into account existing criticism of the dominant

Table 15.1 Different nuclei of meanings regarding the relationship between environmental education and citizenship established by analyzing environmental education Brazilian theses and dissertations in the EArte database

Groups	Subgroups	Characterization
A	A1	Citizenship as a resource to qualify the individual/subject of research.
	35.5%[b]	Concept of citizenship or the process of citizen education is not problematized
48.5%[a]	A2	Indication of political perspectives in EE studies. Concept of citizenship
	13%	or the process of citizenship education is not problematized
B	B1	Indication that the relationship between environmental education and
51.5%	34.5%	citizenship will be studied in depth in the complete text of the research. The citizenship and citizenship education processes are problematized in the text
	B2	Relationship between education and citizenship is a constituent part of
	17%	the questions or objectives of the proposed investigation. EE and citizenship or citizenship education as a research question

[a]Number (percentage) of master and doctoral studies (report research) in each nuclei of meaning groups
[b]Number (percentage) of master and doctoral studies (report research) in each nuclei of meaning subgroups

epistemological paradigm, or critics related to the poor working conditions of those excluded from the current consumerist society. Thus, they point to the meaning of collective action, of diverse segments of society, such as those proposed by environmental movements. Another portion of the articles in this category propose discussing the role of developing social awareness about the process of waste reduction as well as other environmental impacts, such as air pollution, water contamination, and deforestation, associated with social justice and democratic sustainability. Finally, it is worth pointing out that eight of those studies refer to critical theory and to dialectical materialism and consider the relationship between environmental problems and the social, cultural, political, and economic dimensions of society.

In the second group of studies (Group B, Table 15.1), there are abstracts with indicators of problematization of the relationship between EE and citizenship (51.5% or *n*= 262). In this group, there are two subgroups of studies: Subgroup B1 (34.5% or *n*= 176) contains evidence that the relationship proposed was going to be further developed in the complete text of the research. Notably, seven studies clearly associate the citizenship process with other indicators of the political dimension of EE. For example, citizen/citizenship was directly associated with political participation, social and political commitment, social praxis, political education, democracy, community level of governance, shared management, and social and environmental justice, among others. The process of social and economic transformation or the transformation of economic systems and consumerism were also stressed.

The second subgroup of this set (17 % or *n*= 86) is characterized by abstracts where the relationship between EE and the concept or process of citizenship is problematized as a research question. That is, this relationship is considered to be a research problem.

15.4 Education and Citizenship: Challenges and Perspectives in Light of the Environmentalist Discourse

The data analyzed enable us to address some counterpoints with the national and international literature that explores the relationship between education/environmental education and citizenship.

The first point that we would like to raise is the fact that our research results have made us alert of the current risk of trivialization and emptying of the concept of citizenship and of the political dimension of EE discourses. This risk is concrete when, in our research or pedagogical activities, the intrinsic relationship between education/environmental education and citizenship is taken a priori, without a proper clarification of its meaning (Santos et al. 2014). It is possible that the silencing of researchers and educators concerning the meanings attributed to the relationship between education and citizenship means the veiled adoption of a concept of citizenship of liberal inspiration. In this perspective, the central concern is for the individual rights, desires, or life projects as citizenship is primarily seen as a legal statute (Bannell 2010). In addition, it suggests that EE research and pedagogical discourses are very close to those in which the presumed subject–citizens are not real people. As Stauffer and Rodrigues (2011) contend, in this case, subjects "are abstracted from their more entrenched interests, from their social conditions and from the ideological disputes present in society in order to cooperate with the concepts of justice and equity" (p. 36). Thus, it seems justifiable to accept Valle's (2010) proposal to consider our reality and concrete life conditions, making explicit not only what is meant by the word citizen in our countries today but, above all, what we believe any citizen should be.

Another challenging point that we face in the discourse on the role of education/environmental education that has been constructed by researchers and educators lies in the necessary reaction to the influence of neoliberal policies. Many educators have made efforts to make the links between the neoliberal trend and education/environmental education explicit in different contexts (see Hursh et al. 2015).

The neoliberal perspective imposes the principles of private and managerial initiatives on schools, which directly influences not only the aspects of administration and management of the educational system but also curriculum conceptions, political and pedagogical projects, teaching and learning, assessment, and teacher training, among others (Hursh et al. 2015). This tendency brings great difficulties in terms of planning and developing more open views to pedagogical practices, such as interdisciplinary practices. In other words, the neoliberal trend is being assimilated in such a way and to such an extent that we do not realize how much it restricts our thoughts, feelings, and our practices as educators and researchers (Hursh et al. 2015).

It is necessary to point out that in a significant number of Brazilian theses and dissertations on EE analyzed (approximately 300 documents) as well as in different papers and authors, some indicators clearly point to what might be called a reaction to this neoliberal perspective of citizenship. That is, there are signs of reaction at attempts to construct false universalities supported by ideologies that mobilize interests of dominant hegemonic groups.

In all the literature to which we have referred, it appears that there is a certain emphasis to the possible relationship between the process of construction of environmental citizenship and the principles of environmental sustainability. In the case of theses and dissertations carried out specifically in Brazil, regardless of the group to which the research was associated, the terms "sustainable development" and "sustainability" are frequently used. However, the analysis of this set of research reports that we have carried out shows that there is also evidence that researchers are not all talking about the same kind of citizenship, the same kind of EE, or sustainable development or sustainability. This condition raises important issues for this debate. Hursh et al. (2015) explore the proposition by McKenzie et al. (Apud Hursh et. al. 2015) which, when examining the neoliberal policies of mobility, suggests that sustainability should be understood as a "'vehicular idea' in order to understand how the three cornerstones of sustainability are differentially contextualized" (p. 310). What interests us in this observation is the understanding of the variety of meanings attributed and constructed, in different contexts, about the idea of sustainability.

To a certain extent, this variety of meanings about the idea of sustainability could be a chance to understand that, even in contexts of globalization, particularities of environmental policies can generate distinct relationships between education and environmental issues. What we should consider, according to Hursh et al. (2015), are the various strategies of industrial capitalism in transforming the ideal of sustainability into eco-business or market ecology – i.e., the environment transforms into commodities as other possibilities to make profit. In such cases, the concept of sustainability is useful for creating values, for producing profit, always in the name of environmentalism and education.

15.5 Environmental Education and Citizenship: A Conflictive Field

More than an attempt to systematize and present conclusive statements, our intention is to finalize this chapter by emphasizing some provisional understandings about the historical and complex relationship between environmental education and citizenship.

Although our analysis is preliminary, it allows us to recognize that various researchers are taking part in a debate that is open to multiple ideological influences and complex meaning-making processes. Some indicators that we have systematized in this chapter clearly point to what might be called a reaction to the hegemony of the so-called modern and liberal concept of citizenship. Reactions against attempts to construct false universalities, supported by ideologies that mobilize interests of hegemonic groups, are clearly present in some of the theses and dissertations abstracts analyzed.

Our data also revealed how susceptible we are to generalizations and to decontextualized universalizations, supporting conservative or pragmatic perspectives in which the ideal of citizenship is founded on principles of liberal or neoliberal inspiration. This is the reality we face when we engage in environmental education: a conflictive field involving diverse interests, some of which are hegemonic and historically resistant to social changes. There is no room for naivety or poorly coordinated actions (Carvalho 2010).

As researchers and educators, it is important to recognize that we are involved in a context that is full of conflicting ideological interests. This reality, in turn, influences the formulation and the implementation of public policies in education/environmental education as well as knowledge production and dissemination.

Stauffer and Rodrigues' proposal (2011) is very significant when considering the contradictory social and historical nature of education. According to those authors, education is marked by materials and concrete conditions of the society to which it belongs, such as economic model, cultural traditions, or political experiences. It is also inserted in a context of tension that determines the struggles for democratization. The authors' proposition is to think not about education for citizenship, but about education of the citizen, who is seen as a real subject, "present, marked by socio–historical conditions, but not determined by these, or bound to these" (pp. 43–44).

Considering this proposal as a possibility to be constructed, new challenges are posed to associate concrete subject/subjects with the process of citizenship education. This project, which aims to construct autonomy and freedom, presupposes the articulation with other aspects of political practice. In Jacobi's proposals (2003), the idea is to construct a citizen who is a "founder of rights in order to open up new spaces for social and political participation" (p. 315), articulated to a context of constructing a democratic culture (Lima 2011).

Finally, it seems plausible to consider, as proposed by Hursh et al. (2015), that the field of education is always open to problematizing and questioning. If we agree with this statement, then the citizenship education process, always open to new questions, invites us to reflect and to struggle in contexts of broad hegemony of neoliberal ideologies.

Questions

Discussion questions that could help instructors and students to engage in meaningful conversation about the ideas presented in this chapter.

1. What, in your view, are the most significant characteristics of a citizen in our contemporary world? Are these characteristics universal or are there different perspectives of the characteristics or attributes of a citizen?
2. To what degree is the educational process required in the development of citizenship construction? In other words, what are the possibilities and limits of education regarding the process of citizenship construction?
3. Select three papers that in some way relate environmental education to the process of citizenship construction, published in different countries, and systematize the main ideas regarding citizenship conveyed by these papers

4. Compare the ideas that you can get from these papers with the ideas conveyed in the Brazilian journals systematized in this paper that you have read. Can you see any kind of approximation or differences between these papers from different countries?
5. What environmental education practices would you consider to be more consistent with your ideas and/or ideals of citizenship education?

References

Arbués, E. (2014). Civic education in Europe: Pedagogic challenge versus social reality. *Sociology Mind, 4*(3), 226–232.

Bakhtin, M. M. (2010). *Estética da Criação Verbal*. São Paulo: WMF Martins Fontes.

Bannell, R. I. (2010). Pluralismo, identidade e razão: formação para a cidadania e a filosofia política contemporânea. In A. J. Peixoto (Ed.), *Filosofia, educação e cidadania* (pp. 155–207). Campinas: Editora Alínea.

Bannell, R. I., Prata, R. V., & Fenerich, C. (Eds.). (2011). *Educação para a cidadania e os limites do liberalismo*. Rio de Janeiro: 7 Letras.

Carvalho, L. M. (2010). Que Educação Ambiental desejamos? *Revista Ciências em Foco, 1*(3), 1–22.

Conrado, D. M., Nunes–Neto, N. F., & El–Hani, C. N. (2014). Aprendizagem baseada em problemas (ABP) na educação científica como estratégia para formação do cidadão socioambientalmente responsável. *Revista Brasileira de Pesquisa em Educação em Ciência, 14*(2), 77–87.

Cosenza, A., Freire, L. M., Espinet, M., & Martins, I. (2014). Relações entre justiça ambiental, ensino de ciências e cidadania em construções discursivas docentes. *Revista Brasileira Pesquisa em Educação em Ciências, 14*(2), 89–98.

Dimick, A. S. (2015). Supporting youth to develop environmental citizenship within/against a neoliberal context. *Environmental Education Research, 21*(3), 390–402. https://doi.org/10.1080/13504622.2014.994164.

Ferreira, N. S. A. (2002). As pesquisas denominadas "estado da arte". *Educação & Sociedade, 23*(79), 257–272.

Gallo, S. (2010). Filosofia, educação e cidadania. In A. J. Peixoto (Ed.), *Filosofia, educação e cidadania* (pp. 133–153). Campinas: Editora Alínea.

Higuchi, M. I. G., & Moreira Júnior, W. (2009). Educação Ambiental e movimentos sociais: espaços paralelos ou compartilhados? *Pesquisa em Educação Ambiental, 4*(2), 165–174.

Hursh, D., Henderson, J., & Greenwood, D. (2015). Environmental education in a neoliberal climate. *Environmental Education Research, 21*(3), 299–318. https://doi.org/10.1080/13504622.2015.1018141.

Jacobi, P. R. (2003). Espaços públicos e práticas participativas na gestão do meio Ambiente no Brasil. *Sociedade e Estado, 18*(1/2), 137–154.

Jacobi, P. R. (2005). Educação Ambiental: o desafio da construção de um pensamento crítico, complexo e reflexivo. *Educação e Pesquisa, 31*(2), 233–250.

Lima, G. F. C. (2011). Crise ambiental, educação e cidadania: Os desafios da sustentabilidade emancipatória. In C. F. B. Loureiro, P. P. Layrargues, & R. S. Castro (Eds.), *Educação ambiental: Repensando o espaço da cidadania* (5th ed., pp. 115–148). São Paulo: Cortez.

Loureiro, C. F. B. (2011). Educação Ambiental e movimentos sociais na construção da cidadania ecológica e planetária. In C. F. B. Loureiro, P. P. Layrargues, & R. S. Castro (Eds.), *Educação ambiental: Repensando o espaço da cidadania* (5th ed., pp. 73–104). São Paulo: Cortez.

Martins, I. (2006). Dados como diálogo: construindo dados a partir de registros de observação de interações discursivas em salas de aula de ciências. In F. M. T. Santos & I. M. Greca (Eds.), *A Pesquisa em Ensino de Ciências no Brasil e suas Metodologias* (pp. 297–321). Ijuí: Ed. UNIJUÍ.

Payne, P. G. (2009). Framing research: conceptualizing, contextualizing, representation, legitimization. *Pesquisa em Educação Ambiental, 4*(2), 49–77.

Peixoto, A. J. (2010). *Filosofia, educação e cidadania* (3rd ed.). Campinas: Editora Alínea.

Reis, G., & Oliveira, A. (2014). Environmental discourses in science education: Contributions to democracy, citizenship and social justice. *Revista Brasileira Pesquisa em Educação em Ciências, 14*(2), 9–26.

Santos, W. L. P., Carvalho, L. M., & Levinson, R. (2014). The political dimension of environmental education into research of Brazilian journals of science education. *Revista Brasileira Pesquisa em Educação em Ciências, 14*(2), 199–213.

Scott, W. (2011). Sustainable schools and the exercising of responsible citizenship – A review essay. *Environmental Education Research, 17*(3), 409–423. https://doi.org/10.1080/13504622.2010.535724.

Stahelin, N., Accioly, I., & Sánchez, C. (2015). The promise and peril of the state in neoliberal times: Implications for the critical environmental education movement in Brazil. *Environmental Education Research, 21*(3), 433–446. https://doi.org/10.1080/13504622.2014.994167.

Stauffer, A. B., & Rodrigues, R. L. (2011). O liberalismo político de Rawls e a educação: qual a perspectiva para a formação do cidadão? In R. I. Bannell, R. V. Prata, & C. Fenerich (Eds.), *Educação para a cidadania e os limites do liberalismo*. Rio de Janeiro: 7 Letras.

Valle, L. A. B. (2010). Modelos de cidadania e discursos sobre educação. In A. J. Peixoto (Ed.), *Filosofia, educação e cidadania (3rd ed)* (pp. 213–235). Campinas: Editora Alínea.

Wolmann, E. M., & Braibante, M. E. F. (2014). Utilizando a elaboração de folders para a construção da cidadania com estudantes do Ensino Médio. *Revista Brasileira Pesquisa em Educação em Ciências, 14*(2), 265–278.

Chapter 16
Black Earth Green Moon Mama Allpa: Polyphonic Moments from Temple to Tambo

Patricia Palulis

Abstract During the last month of an academic leave from the University of Ottawa, I take a detour and attend a summer institute at the Sachamama Center for BioCultural Regeneration (SCBR) located in the Peruvian High Amazon and under the direction of Frédérique Apffel-Marglin. It was organized by scholars Peter Cole and Pat O'Riley from the University of British Columbia. At Sachamama, I was drawn into ecopedagogical responsibility, working against the grain of the progress narrative. Returning home, I work with my students to intensify the provocations at the margins of Academia. Storytelling and mapping help us to open generative possibilities for eco-literacies and geo-literacies as we work with the tricksters, Coyote and Raven, to decolonize and indigenize teacher education.

Qintipa kirunwanshina	*Con el pico de un colibri*	With the beak of a hummingbird
kay takikunata	*estos cantos*	I offer you these
kushikichi	*les ofrezco*	Songs… (Marcos 2013, p.3)

This chapter opens with a poem brought back from the BioCultural Regeneration Institute at the Sachamama Center in Lamas, Peru. With the beak of a hummingbird, Bárbara Rodrigues Marcos draws exquisite attention to Grandmother Soil and brings us rich cadences of poetic wisdom. I thank Abby Corbett for the English translation.

Yana allpanchikkuna	*Nuestra tierra negra*	Our black soil
kashkan	*está surgiendo*	is emerging
rikuriykan.	*nuevamente.*	once again.
Awila Allpa	*La Abuela Tierra*	Grandmother Soil
kashkan	*otra vez*	lives
kawsan.	*vive.*	again. (Marcos 2013 p.5)

Three languages work together singing their stories allima muy bien excellent. Storytelling at work in the High Amazon shelters me upon my return to the Land of Academia. Reading Bhabha (2007), I am drawn to his contention: "Globalization…

P. Palulis (✉)
University of Ottawa, 145 Jean-Jacques Lussier, Ottawa, ON K1N 6N5, Canada
e-mail: ppalulis@uottawa.ca

G. Reis, J. Scott (eds.), *International Perspectives on the Theory and Practice of Environmental Education: A Reader*, Environmental Discourses in Science Education 3, https://doi.org/10.1007/978-3-319-67732-3_16

must always begin at home" (p. 9). Working at home within the tensioned spaces of local/global resonances and dissonances, I am yearning for another kind of pedagogy. Running away from home on a sabbatical leave, space opens into the unknown. During the last month of my sabbatical leave, I take a detour from my plans and attend a summer institute at the Sachamama Centre for BioCultural Regeneration (SCBR) under the direction of Frédérique Apffel-Marglin. The summer institute was being taught by scholars Peter Cole and Pat O'Riley from the University of British Columbia. We learn about biocultural regeneration of pre-Columbian black earth, the phases of the moon, sowing and harvest times and the sacred rituals of indigenous traditions. At Sachamama I was drawn into an ecopedagogical responsibility, working against the progress narrative. Returning home, the work of unlearning begins in the praxis of everyday life. I thank Peter Cole for the provocative transformation of itinerary into itineRANT.

16.1 Provocations and Recollections in the Nation's Capital

Stories are not research. I have been told. A mandated not-ness resounding with authority. The arrogance of the voice assuming such authority to speak on behalf of stories. I read this not-ness as a posting of a sign, a signature, a signatory for detours – exquisite seekings for subversive escapes – for indwelling on fault lines. Donna Haraway (1997) contends that numbers are just another way to tell a story. And so I return to Haraway for backup: "Not only is no language, including mathematics, ever free of troping; not only is facticity always saturated with metaphoricity; but also, any sustained account of the world is dense with storytelling" (p. 64). In tracing the entanglements of recollections, the reader is invited to engage with fragments of storytelling that work to destabilize the linearity of temporality and the spatiality of geography.

Entering into my textual travels dense with storytelling, I draw from Michel de Certeau (1998) on the signifying practice as "the itinerary of rupture, the implementation of the unpredictable, the 'poetics' of play, the disorganization of conventional arrangements…injecting an *internal semantic shock*, an explosion that disrupts the dominant social order…of signifiers" (p. 33). At SCBR, we enter into the resistance of the progress narrative – an itinerary of rupture.

As darkness descends, the night opens to possibilities of becoming dense with storytelling. We enter the small tambo – into the interior darkness – sitting on low wooden benches one beside the other; women on one side, men on the other side. The shaman enters and remains standing by the door. I watch intently as the first person is smudged by the shaman. The darkness is broken only by the sparks of the flames as the mapacho lights up and flickers out – the shaman chanting and smudging us one by one in a sacred ritual to protect us. And then it is my turn – an invocation into the spirituality of the sacred tradition. Rituals become an integral part of our experience at Sachamama.

16.1.1 Ecstatic Gifting and Splashing Speech

In the small library collection at Sachamama, I find a familiar author in David Abram (1996). We are indwelling in the titles in the cadence of his spell of the sensuous:

> As the splashing speech of the rivers is silenced by more and more dams, as we drive more and more of the land's wild voices into the oblivion of extinction, our own languages become increasingly impoverished and weightless, progressively emptied of their earthly resonance. (p. 86)

If we cannot speak in a chorus with the "splashing speech of the rivers", then are we not complicit in the emptiness and impoverishment? I read Abram while heeding the tricksters Peter Cole and Pat O'Riley (2012) as they teach: "Hands up and down to earth" – retellings and remappings – a poetics of serious play. Earth and water. In the attention to translating water, David Jardine's soundwalks echo and accompany me here to the biocultural institute. I become attentive to his wisdom. Words are not stand-ins, he cautions.

> Words that bespeak the sound of water are meant to make it present, to show it off, to lead us to it and offer us up to its ways, not to stand in front of it and block our way. They are not substitutes but rather heralds of the arrival of the thing. (Jardine 2008 p. 17)

So entrenched in words and books and authors, I can sense the sensuous as words disappear into droplets of water. I draw out a thread from Apffel-Marglin (2011) reading Linda Hogan on the spirit of the gift "standing in the rain ecstatic with what is offered" (p. 64). It rained almost every day in Sachamama. Short bursts of a downpour as ecstatic giftings. As watermarks blur the words. Resisting the authority of canonical words. The sign carriers of Eurocentric authority. In this small library, I am introduced to the work of Karen Barad (2007). Entanglements reintroduced through the work of Barad became my working space – entangled in the smudging and chanting of the shaman – attending to a sonata in the chorus of frogs and cicadas, indwelling within the ecological context of Sachamama within the earthly resonances of Abram.

16.1.2 Quiet Cacophonies and Lively Tensions

Reading (with) the lively tensions of Barad (2012), words become cacophonies and symphonies:

> Virtual particles are not in the void but of the void. They are on the razor edge of non-being. The void is a lively tension, a desiring orientation toward being/becoming. The vacuum is flush with yearning, bursting with innumerable imaginings of what could be. The quiet cacophony of different frequencies, pitches, tempos, melodies, noises, pentatonic scales, cries, blasts, sirens, sighs, syncopations, quarter tones, allegros, ragas, bebops, hiphops, whimpers, whines, screams, are threaded through the silence, ready to erupt, but

> simultaneously crosscut by a disruption, dissipating, dispersing the would–be sound into non/being, an indeterminate symphony of voices. (p. 13)

Entangled in writings, I attend to Ted Aoki (1993) who taught me how to listen otherwise, to dwell elsewhere:

> What kind of a place is this? A place where there is room for words like *humour*, *human*, *humus*, *humility* to live together. In such a place to be humiliated is to be reminded that we are communally ecologic, that the rhythmic measures of living on Earth come forth polyphonically in *humour* and *human* and *humus* and *humility*. (p. 300)

It was here in the Peruvian High Amazon that his words took root. My humiliation becomes a space for becoming in tension. Aoki extended an exquisite invitation with his provocative assignment: to write a paper/not paper. He invited us to experiment with writing as arche-texture. To risk in-citing, ex-citing and re-citing. I was soon entering Cixous' (1997) three schools of writing: the school of the dead, the school of dreams and the school of roots. I have been working my way through her schools of writing. I am attending, finally, the school of roots. On assignment. At Sachamama in the Peruvian High Amazon. Cixous' three schools of writing begin with an acknowledgement of her affinities. The roots are entangled in writings in watermarks in elsewhere spaces: "Elsewhere, outside, birds, women, and writing gather" (p. 113). Cixous becomes an intertextual tying with the songs of Marcos, with the beak of a hummingbird. At home, my writing dense with storytelling, words emerge from out of the humus of my notes. With humility, I am relearning unlearning. Archival ephemera as soundings out of place held in mind. Recollections. Body bitten.

> Shaman chanting icaros, smoking mapacho, smudging, elders drumming, full moon rising, frogs croaking, cicadas chorus, geckos falling, bat poop droppings, fireflies flickering in the night trees, parrots shrieking, turkey vulchers circling, raindrops, heartbeats, pulses throbbing, chica brewing, roosters crowing, horses grazing, dogs barking, piglets rutting, bullet ants: beware. Bugs everyWhere. Tarantulas. (Palulis 2013 ItineRANT fieldnotes)

16.1.3 Polyphonic Moments and Geo-metrons

Escaping into another kind of pedagogy, I attend to the resonances of Aoki's (1991) geo-metrons: "Let's listen. Geometry is geo–metron. 'Geo' is 'eco'; it is earth, this earth on which we dwell, this earth whose humus nourishes us" (p. 374). Aoki traces polyphonic movements: "Earth, measure, temple, mouth, echoes, to speak to say – these are the polyphonic strands of poetry" (p. 375). From a temple to a tambo – a polyphonic movement, a topographical drift, a cartographic shift. A visit to the Kichwa-Lamista village of Shukshuyaku brought us closer to this earth whose humus nourishes us. We were taught the biocultural skills of the indigenous people who use the local habitat to build shelter and to provide nourishment. The people offer hospitality. They welcome us. They teach us. So that we can begin to unlearn and relearn. Working against the grain of the progress narrative. We enter the village of Shukshuyaku. A tambo has been set up for us as sleeping quarters.

We enter with our headlamps, flashlights, cellphones, iphones, guitars, scrabble games, card games, water bottles, wet wipes, toilet paper, mosquito netting, sleeping bags, deep insect repellent, after bite lotion. And bottles of medication for every possible ill. Slowly, some of us learn to inhabit a different rhythm and to appreciate the cycles of sun and moon. The children catching small frogs and dancing with fireflies in the darkness of the evening. The dark comes with the setting of the sun. There is no electricity. We are grinding corn for the chicha – for the rituals of the solstice – we pluck chicken feathers – we grind coffee beans – we are preparing for the feast. (Palulis 2013 ItineRANT fieldnotes)

I participated on the periphery. I am an old woman taking a short turn at grinding corn, rice and coffee beans. I am plucking chicken feathers. Cold water showers inside at Sachamama become cold water showers outside at Shukshuyaku. Cixous (1997) cautions us: "Since we are shaped by years and years of all kinds of experiences and education, we must travel through all sorts of places that are not necessarily pleasant to get there: our own marshes, our own mud" (p. 119). An outhouse was constructed for our use. I slip slide down a slope where the clay has been softened by rain. Climbing down from the outhouse, I slip on the wet grass and slide down the slipperiness of the slope of a clay ridge landing in front of the tambo in Shukshuyaku. Humbled by embodied entanglement, I find the etymological humour in Aoki's humus. Marshes and mud or grasses and clay. I find humour in my humiliation as the villagers look on. By the time the second group arrives, the villagers have constructed a handrail down the slope of the trail from the outhouse to the living quarters in the tambo. My body bitten by a multitude of winged and wingless creatures, Professora Ida Gonzales Flores provides a remedy derived from local plants. I cover my bites with her soothing remedy. I carry my scars home with me. Bitten with body marks with watermarks. We learn to cover exposed skin. To wear our knee-high rubber boots.

16.1.4 Signs and Sign Carriers

Reading Apffel-Marglin (2011) reading Mignolo, I am confronted with my complicity as a teacher of language and literacies:

> As the remarkable scholarship of Walter Mignolo on the tight link between the wielding of alphabetic literacy as a tool of colonial domination in the conquest and administration of Latin America by the Spaniards abundantly demonstrates, alphabetic literacy has marginalized all other forms of reading. Mignolo contends that there is no language that does not come also with reading. However, how such reading is done varies greatly among languages, from the reading of signs such as those left by animals, or the configuration of constellations, to ideograms or other non–alphabetic visuals. (p. 190)

I am confronted again and again with the conditions of my (il)literacies. I return to Mignolo (2006) who provokes us with the materiality of reading and writing cultures. As I step out of papered tenancies, I want to learn with Mignolo to read the signs and the sign carriers that I encounter here at Sachamama and Shukshuyaku.

Another kind of reading. One that works at resisting the progress narrative. Another kind of hospitality. Appreciative of our coming to learn and not to teach. The signs of soundscapes, river walks. A Jardinian Kerplunk!

I never read alone. Reading Abram and Jardine, I return to reading Wanda Hurren (2003) reading Jardine's (2000) collection of ecopedagogical essays in "Under the Tough Old Stars". A reader enters into river spray. Hurren describes her emergence from the text:

> Glistening with river spray. Such a moist, scholarly text.... His writing is a performance of how place and self mingle in curricular theorizing. His notions of ecopedagogy are filled with autobiographical accounts of places in his world where he is inspired, where he learns to listen, where he watches himself being watched, at the river edge or in a stand of aspen. (p. 114)

Cixous (1993) inscribes us as "living particles, fireflies in the world" while "around us resounds an enormous concert of noise–and–rumor–producing machines, creating a din" (p. 6). Climbing the hill towards the town of Lamas, we enter the din of the concert (dis)concerting. As I read into the spaces in-between, Aoki's geometrons resound with Cixous and Barad. Always a clustering. Always a chorus. The entanglements keep winding around each other as writers are greeted and hosted in someone else's text. Words create spaces for meeting places.

16.1.5 Water Marks and Water Texts

Jardine (2008) makes reference to a "water text" composed by his son – soundings of water through soundings of words. Alone with the rains of Sachamama, I bring out my video camera when everyone is away for the second village visitation. I needed solace and silence to slow down – to heed the cacophonies of the silence of the void. I am alone here on the balcony listening to the geo-metrons of the falling rain. The silence of my listening resonating within the geo-metrons.

The rain trickles and trebles on brilliant pink papery petals of bougainvillea blossoms. Rain falls like poetry on the paperies of petals clinging to their branches, holding firm despite the torrents of rainfall. The soundings as water text shelter me in these verdant cacophonies of silence. Seeking returns to the void. Becomings of the void. Becomings of the world.

The emptiness of the signifying crack in Aoki's paper/not paper invokes what Barad (2012) articulates as an "infinite plentitude of openness" (p.16). I encounter Aokian geo-metrons in the rainforest. As I work nearby with Jardine (2008) translating water as I store my words and story my world with Carl Leggo (1995). In the worksite that we inhabit. A worksite that nourishes us as writers. At school with Cixous (1993): "We must work. The earth of writing. To the point of becoming the earth. Humble work..." (p. 156). Here in the High Amazon rainforest, her words take on new resonance.

16.2 Raining into the Earth

I become rain. I am a waterfall. I am raining. I am raining into the earth. My body responding to intense heat and humidity. Becoming the earth. We have walked uphill to gather sacks of microorganisms for the preparation of dark earth. I find a tree stump to sit on holding a branch to keep the winged and wingless at a distance watching the labouring bodies gathering the bounty of microorganisms from the humus. I am humiliated by my inability to contribute to the labour. At home, I begin to write my story, and words emerge from out of the humus. Embodiment. Entanglement. Overflowing. Halted.

For David Jardine (2008), "Where water sounds, water breathes. Hearing the sound of water is hearing the breathing (aeration) of water …to hear the breathing of water is to be the one who breathes" (p. 15). I want to breathe.

16.2.1 A Gecko on My Windowpane

I awake in the early morning light to see a gecko on my windowpane. A good sign, I am told. A sign carrier. It is the last day of my academic leave. And I am leaving Sachamama today. I ran away from home repeatedly during my academic leave. Returning unhomed and unhomely. Now I am seeking returns with the remains of the leave. I am ready to move on attending to the wonder and the resonance of the geo-metron. Awaiting new growth to emerge from the humus and the humiliation of lived experience in the Land of Academia. To seek subversive passageways for smuggling Aoki's geo-metrons onto campus, into curriculum, onto the screen and off again, into the corridors, onto the grounds. At work in the fault lines. The screams that dwell within are released through the shrieks of the parrots. Threaded through the silence – returning with Coyote and Raven nearby – the tricksters always already cocooning me in ecological spaces of indigeneity. Reading Cole and O'Riley (2012), I am returned to reading Henry David Thoreau with them: "Education… makes a straight–cut ditch of a free, meandering brook" (p.18).

When we drop words on the page, we leave our trail markings. I ran away from a straight-cut ditch, and now I am going home to seek a meandering brook. To become in the seeking of the meandering brook. To invite my students to follow the meandering brook – until I begin to understand that we are of the meandering brook. We meander as we chase after our affinities.

I begin my indwelling with this stranger [not] at home. I am not studying Sachamama. I am reading and rereading the marks and the traces that Sachamama has left on me. Semantic shock – recoding the aftershocks – here on papered blanks. Aoki's assignment lives on: to write a paper/not paper. An interminable paper that cannot stop writing. I am always on assignment. Barad (2012) evokes the possibilities of the blank page: "The blank page teeming with the desires of would–be traces of every symbol, equation, word, book, library, punctuation mark, vowel, diagram, scribble, inscription, graphic, letter, inkblot, as they yearn toward expression. A jubilation of emptiness" (p.13).

On assignment at Sachamama, the words begin to stir. At home, I want to invite my students to enter a blank page teeming with a jubilation of emptiness. My intentions become increasingly inspirited following Sachamama. The students often exceed my intentions, my expectations and my ability to imagine. As we work with body biographies, the students become bricoleurs as they work with the materials at hand.

Returning, I seek exquisite attention to a jubilation of emptiness – awaiting the traces as they yearn towards expression. And sometimes from a jubilation of emptiness – extending my hands to receive what is offered as a yearning of joyful expression. I am revisioning what it means to receive the gift of an assignment. What are tears for? Sometimes they perform for words when words are silenced or fearful. Words rain. They become a rainfall in deforested areas. Teaching in rooms with no windows. Enclosed in curricular templates.

De la luna – allima – full moon. It is the evening before my last class. I am walking home in the dark cold night pensive, moody, tired when I glance up at the sky encountering la luna in circular splendour. Startled by my recollections. The drumming of the elders evokes a soundscape to greet the moon – from my position in the terroir of an icy Sweetland Street in Ottawa. Spring is reluctant to arrive. La luna returns me to the month of June in Sachamama and the drumming of the elders for the sacred ritual of the full moon. I return to the poetry of Barbara Rodrigues Marcos:

Llipyay llipyay	*Brilla, brilla,*	Shine, Shine,
mama killa!	*Madre Luna!*	Mother Moon! (Marcos 2013, p. 15)

16.2.2 Field Notes as Homework

Barad (2012) contends: "A field is something that has a physical quantity associated with every point in spacetime" (p. 10). My recollections vibrate with the soundscapes of elders drumming and chanting. Attending to Barad, flow continues in space-time: "If a drummer now taps the drumhead, it vibrates, and waves of energy flow outward from where it is tapped" (p. 10).

We become entangled in the texts of the writers with whom we have an affinity. We are always studying with one another riding the waves of energy. A hummingbird flutters from the stilled pages of a poem as I await the long-delayed arrival of spring in the nation's capital. I cannot write alone on a blank page. And so I retain my mentoring scribes in a polyphony of sound bytes. If I can fan a tiny spark of that yearning towards expression, if I can be a witness to a tiny spark igniting from a jubilation of emptiness, then, perhaps, I can ride the wave of energy.

16.2.3 Mappings and Storied Yarns

Returning to Bhabha, the homework begins again. How do I bring home a global experience and create with my students the earthly resonance for a yearning towards expression? In paths of resistance to the progress narrative? Apffel-Marglin (2011) reading (with) Barad contends that materiality and discursive practices are always entangled rendering representation an impossibility. How can we acknowledge entanglements in pedagogical practices? Let go of our binaries and hierarchies? Open to the soundings of silences? Seeking detours from hegemonic structures and strictures opening to regeneration of indigenous pedagogies.

Reading Lucy Lippard (1997), I am attending to the lure of the local. Lippard contends that "Each time we enter a new place, we become one of the ingredients of an existing hybridity, which is really what all 'local places' consist of" (p. 6). I return to her citing of Doug Aberley (2003) on home-made maps: "Reinhabitants will not only learn to put maps on paper, maps will also be sung, chanted, stitched and woven, told in stories, and danced across fire–lit skies" (p. 5). With Lippard, I am lured into a sensing of place, wanting to weave myself into the experiences of mapping in the Peruvian High Amazon. At home, hybrid mappings learned from the lure of the local at Sachamama are recreated.

In my teacher education classes, my assignments begin with hometown poems and recycled mappings of home-grown cartographies. I am drawing from long-ago projects seeded and nurtured through the years. We map our worlds with images and words – grounded and not grounded. Curriculum and pedagogy become geometrons. The mappings interwoven with storytelling as nomadic destinations. I return to Hurren (2003) inspired by her curricular collage of auto'geo'carto'graphia.

Having run away from home, I am now returning unhomed. Only the eyes are raining now. I am dropping tears on the words of Hélène Cixous (1998) as I seek refuge in her escaping texts: "And repentance? No repentance" (p. 22). I would run away again. And again. From curriculum documents and from scholarly canons from the authorial dictations. Returning I will inhabit the solitudes, dwell with the spectres, cocooning in silence with strings of letters mapping cartographies of the void. I attend to Barad (2007) and the entanglement of matter and meaning.

16.2.4 Postscripts and River Walks

Sachamama has marked me. I have shifted from the Rideau Canal to the Rideau River, from straight-cut to meandering brook. I live in a small tree house loft surrounded by vined entanglements and sheltered by towering trees from neighbouring yards. I hang bird feeders in the yard and take delight in chickadees arriving to feast on seeds. Sachamama has inhabited me. I am only beginning to dwell on what that might mean in a difficult class with difficult knowledge. In the light of early morning, I begin to water my little garden of jasmine, lavender, bougainvillea, and as I rummage to untangle the vines, I see the blossoming of a passion flower. I take a detour. Tears of anguish dissipate into a jubilation of emptiness. To begin again always somewhere else. Walking down to the Rideau River one encounters river people, those who walk by the river's edge and those who linger on the river wall on the park benches on the grass. Mappings made by footprints on moon walks by the river's edge. River walks to greet a full moon. Ephemeral recollections of the elders chanting and drumming. River people greet one another on the path by the river. Geese, ducks and seagulls settle on rocky perches in the shallow waters of the river. A blue heron.

As Abram (1996) reminds us, "our task…is that of taking up the written word, with all its potency, and patiently, carefully, writing language back into the land" (p. 273). To write language back into the land, we must attend to our whereabouts. How can we learn to work with our students attending to Mignolo's sign carriers, Aoki's geo-metrons, Barad's entanglements, while engaging in serious play with the tricksters, Pat O'Riley and Peter Cole, Coyote and Raven. I want to read the tricksters with Gough (2008) querying how places become pedagogical. I stumble over a preposition as I search for a return to the footprints that do not stop at the river's edge. The river walk is temporarily suspended as the construction of a footbridge over the Rideau River is under way. I venture into the marshes and the mud again seeking new nesting grounds. I had been looking for a meandering brook but now I want to indwell in the geo-metrons of a meandering brook seeking Aokian moment(um)s crossing again to the other side of the canal. I had settled into a backyard loft in the forest. The darkness and isolation of long winter months unsettle me, and I am on the move again seeking the visibility of street lights and neighbourhood. Drawn into the circulation of the city, I am becoming of the void of the world.

16.2.5 A Tremble in the Fault Lines of the Progress Narrative

In the name of progress, the trees around the Faculty of Education are being destroyed to make space for the construction of a new learning centre. I experience an internal semantic shock. My camera quivers as I am recording the shrieking of the chain saws as they do their daunting work, the grinding of branches being eaten alive by the metal jaws of machines and the haunting whispers of tree dust being swept away by a tidy broom. And then there is the silence of the trees! Abram attends to Merleau-Ponty who reminds us: "Language is the very voice of the trees, the waves, and the forest" (cited in Abram 1996, p.86). As I mourn the death of our trees, I mourn even more deeply our (non)response to this destruction – refrains of remorse – what can we do, we have no choice, echoes of (non)resistance to the progress narrative. I am haunted by the (im)possibilities of teaching eco-literacies. Of mapping geo-literacies.

I want the forest before the book,
the abundance of leaves
before the pages.
(Cixous 1998 p. 20)

Questions

Discussion questions that could help instructors and students to engage in meaningful conversation about the ideas presented in this chapter.

1. How can we begin to disrupt the progress narrative as artful activism and environmental responsibility?
2. How can we invite our students to engage in creative homemade mapping activities that encourage place as pedagogy?
3. In what ways might arts-based pedagogies contribute to ecopedagogies?
4. Where can we locate ourselves in the discourse so that we might begin to work at decolonizing and indigenizing academia?
5. How can we encourage storytelling as live(d) experience to share with the community?

References

Aberley, D. (2003). The lure of mapping: An introduction. In D. Aberley (Ed.), *Boundaries of home: Mapping for local empowerment* (pp. 1–7). Gabriola Island: New Society Publishers.

Abram, D. (1996). *The spell of the sensuous: Perception and language in a more than human world*. New York: Pantheon Books.

Aoki, T. (1991). Sonare and videre: A story, three echoes and a lingering note. In W. Pinar & R. Irwin (Eds.), (2005), *Curriculum in a new key: The collected works of Ted T. Aoki* (pp. 367–376). Mahwah: Lawrence Erlbaum.

Aoki, T. (1993). Humiliating the Cartesian ego. In W. Pinar & R. Irwin (Eds.), (2005), *Curriculum in a new key: The collected works of Ted T. Aoki* (pp. 291–301). Mahwah: Lawrence Erlbaum.

Apffel-Marglin, F. (2011). *Subversive spiritualities: How ritual enacted the world*. New York: Oxford University Press.

Barad, K. (2007). *Meeting the universe halfway: Quantum physics and the entanglement of matter and meaning*. Durham: Duke University Press.

Barad, K. (2012). *What is the measure of nothingness? Infinity, virtuality, justice*, 100 Notes, 100 Thoughts: Documenta Series (13) 099 by Hatje Cantz; Bilingual edition (September 30, 2012).

Bhabha, H. (2007). Ethics and aesthetics of globalization: A postcolonial perspective. In*The urgency of theory*. Manchester: Carcanet Press & Fundação Calouste Gulbenkian.

Cixous, H. (1993). *Three steps on the ladder of writing* (S. Cornell & S. Sellers, Trans.). New York: Columbia University Press.

Cixous, H. (1998). *Stigmata: Escaping texts*. London/New York: Routledge.

Cixous, H., & Calle-Gruber, M. (1997). *Hélène Cixous rootprints: Memory and life writing*. London/New York: Routledge.

Cole, P., & O'Riley, P. (2012). Coyote and Raven put the 'digital' in technology – Hands-up and down to earth. *Transnational Curriculum Inquiry, 9*(2), 18–34.

De Certeau, M. (1998). Propriety. In D. Certeau, M., Giard, L., & Mayol, P. *The practice of everyday life*, 2, 15–34. (T. J. Tomasik, Trans.). Mineapolis: University of Minneapolis Press.

Gough, N. (2008). Ecology, ecocriticism and learning: How do places become 'pedagogical'? *Transnational Curriculum Inquiry, 5*(1), 71–86.

Haraway, D. (1997). *Modest–Witness@Second–Millenium.FemaleMan–Meets–OncoMouse: Feminism and technoscience*. New York: Routledge.

Hurren, W. (2003). Auto'geo'carto'graphia' (A curricular collage). In E. Hasebe-Ludt & W. Hurren (Eds.), *Curriculum intertext: Place/language/pedagogy* (pp. 111–121). New York: Peter Lang.

Jardine, D. (2000). *Under the tough old stars*. Brandon: Foundation for Educational Renewal, Inc.

Jardine, D. (2008). Translating water. *Journal of Curriculum Theorizing, 24*(1), 11–19.
Leggo, C. (1995). Storing the word/storying the world. *English Quarterly, 28*(1), 5–11.
Lippard, L. (1997). *The lure of the local: Senses of place in a multicentered society*. New York: The New Press.
Marcos, B. R. (2013). *Yana allpa, awila allpanchikkuna kashkan rikurin*. Lima: Dunduri.
Mignolo, W. D. (2006). *The darker side of the renaissance: Literacy, territoriality and colonization*. Ann Arbor: University of Michigan Press.

Index

G. Reis, J. Scott (eds.), *International Perspectives on the Theory and Practice of Environmental Education: A Reader*, Environmental Discourses in Science Education 3, https://doi.org/10.1007/978-3-319-67732-3

MIX
Papier aus verantwortungsvollen Quellen
Paper from responsible sources
FSC® C105338

If you have any concerns about our products,
you can contact us on
ProductSafety@springernature.com

In case Publisher is established outside the EU,
the EU authorized representative is:
Springer Nature Customer Service Center GmbH
Europaplatz 3, 69115 Heidelberg, Germany

Printed by Libri Plureos GmbH
in Hamburg, Germany